Benign by Design

ACS SYMPOSIUM SERIES **577**

Benign by Design

Alternative Synthetic Design for Pollution Prevention

Paul T. Anastas, EDITOR
Carol A. Farris, EDITOR
*Office of Pollution Prevention and Toxics
U.S. Environmental Protection Agency*

Developed from a symposium sponsored
by the Division of Environmental Chemistry, Inc.,
at the 206th National Meeting
of the American Chemical Society,
Chicago, Illinois,
August 22–27, 1993

American Chemical Society, Washington, DC 1994

Library of Congress Cataloging-in-Publication Data

Benign by design: alternative synthetic design for pollution prevention / Paul T. Anastas, editor, Carol A. Farris, editor.

p. cm.—(ACS symposium series, ISSN 0097–6156; 577)

"Developed from a symposium sponsored by the Division of Environmental Chemistry, Inc., at the 206th National Meeting of the American Chemical Society, Chicago, Illinois, August 22–27, 1993."

Includes bibliographical references and indexes.

ISBN 0–8412–3053–6

1. Organic compounds—Synthesis—Industrial applications—Congresses. 2. Waste minimization—Congresses.

I. Anastas, Paul T., 1962– . II. Farris, Carol A., 1947– . III. American Chemical Society. Division of Environmental Chemistry, Inc.. IV. American Chemical Society. Meeting (206th: 1993: Chicago, Ill.) V. Series.

TP247.B395 1994
660—dc20   94–38830
                                                             CIP

The paper used in this publication meets the minimum requirements of American National Standard for Information Sciences—Permanence of Paper for Printed Library Materials, ANSI Z39.48–1984. ∞

Copyright © 1994

American Chemical Society

All Rights Reserved. The appearance of the code at the bottom of the first page of each chapter in this volume indicates the copyright owner's consent that reprographic copies of the chapter may be made for personal or internal use or for the personal or internal use of specific clients. This consent is given on the condition, however, that the copier pay the stated per-copy fee through the Copyright Clearance Center, Inc., 27 Congress Street, Salem, MA 01970, for copying beyond that permitted by Sections 107 or 108 of the U.S. Copyright Law. This consent does not extend to copying or transmission by any means—graphic or electronic—for any other purpose, such as for general distribution, for advertising or promotional purposes, for creating a new collective work, for resale, or for information storage and retrieval systems. The copying fee for each chapter is indicated in the code at the bottom of the first page of the chapter.

The citation of trade names and/or names of manufacturers in this publication is not to be construed as an endorsement or as approval by ACS of the commercial products or services referenced herein; nor should the mere reference herein to any drawing, specification, chemical process, or other data be regarded as a license or as a conveyance of any right or permission to the holder, reader, or any other person or corporation, to manufacture, reproduce, use, or sell any patented invention or copyrighted work that may in any way be related thereto. Registered names, trademarks, etc., used in this publication, even without specific indication thereof, are not to be considered unprotected by law.

PRINTED IN THE UNITED STATES OF AMERICA

1994 Advisory Board

ACS Symposium Series

M. Joan Comstock, *Series Editor*

Robert J. Alaimo
Procter & Gamble Pharmaceuticals

Mark Arnold
University of Iowa

David Baker
University of Tennessee

Arindam Bose
Pfizer Central Research

Robert F. Brady, Jr.
Naval Research Laboratory

Margaret A. Cavanaugh
National Science Foundation

Arthur B. Ellis
University of Wisconsin at Madison

Dennis W. Hess
Lehigh University

Hiroshi Ito
IBM Almaden Research Center

Madeleine M. Joullie
University of Pennsylvania

Lawrence P. Klemann
Nabisco Foods Group

Gretchen S. Kohl
Dow-Corning Corporation

Bonnie Lawlor
Institute for Scientific Information

Douglas R. Lloyd
The University of Texas at Austin

Cynthia A. Maryanoff
R. W. Johnson Pharmaceutical
Research Institute

Julius J. Menn
Western Cotton Research Laboratory,
U.S. Department of Agriculture

Roger A. Minear
University of Illinois
at Urbana–Champaign

Vincent Pecoraro
University of Michigan

Marshall Phillips
Delmont Laboratories

George W. Roberts
North Carolina State University

A. Truman Schwartz
Macalaster College

John R. Shapley
University of Illinois
at Urbana–Champaign

L. Somasundaram
DuPont

Michael D. Taylor
Parke-Davis Pharmaceutical Research

Peter Willett
University of Sheffield (England)

Foreword

THE ACS SYMPOSIUM SERIES was first published in 1974 to provide a mechanism for publishing symposia quickly in book form. The purpose of this series is to publish comprehensive books developed from symposia, which are usually "snapshots in time" of the current research being done on a topic, plus some review material on the topic. For this reason, it is necessary that the papers be published as quickly as possible.

Before a symposium-based book is put under contract, the proposed table of contents is reviewed for appropriateness to the topic and for comprehensiveness of the collection. Some papers are excluded at this point, and others are added to round out the scope of the volume. In addition, a draft of each paper is peer-reviewed prior to final acceptance or rejection. This anonymous review process is supervised by the organizer(s) of the symposium, who become the editor(s) of the book. The authors then revise their papers according to the recommendations of both the reviewers and the editors, prepare camera-ready copy, and submit the final papers to the editors, who check that all necessary revisions have been made.

As a rule, only original research papers and original review papers are included in the volumes. Verbatim reproductions of previously published papers are not accepted.

M. Joan Comstock
Series Editor

Contents

Preface ix

In Memoriam xi

OVERVIEW

1. Benign by Design Chemistry 2
   Paul T. Anastas

2. Environmentally Benign Chemical Synthesis and Processing
   for the Economy and the Environment 23
   Kenneth G. Hancock and Margaret A. Cavanaugh

BENIGN CHEMISTRY: RESEARCH

3. Microbial Biocatalysis: Synthesis of Adipic Acid
   from D-Glucose 32
   Karen M. Draths and John W. Frost

4. Mechanistic Study of a Catalytic Process for Carbonylation
   of Nitroaromatic Compounds: Developing Alternatives
   for Use of Phosgene 46
   Wayne L. Gladfelter and Jerry D. Gargulak

5. Preparative Reactions Using Visible Light: High Yields
   from Pseudoelectrochemical Transformation 64
   Gary A. Epling and Qingxi Wang

6. A Photochemical Alternative to the Friedel–Crafts Reaction 76
   George A. Kraus, Masayuki Kirihara, and Yusheng Wu

7. Mn(III)-Mediated Electrochemical Oxidative Free-Radical
   Cyclizations 84
   Barry B. Snider and Bridget A. McCarthy

8. Supercritical Carbon Dioxide as a Medium for Conducting
   Free-Radical Reactions 98
   James M. Tanko, Joseph F. Blackert, and Mitra Sadeghipour

9. **The University of California–Los Angeles Styrene Process** 114
   Orville L. Chapman

   BENIGN CHEMISTRY: INDUSTRIAL APPLICATIONS

10. **Generation of Urethanes and Isocyanates from Amines and Carbon Dioxide** 122
    Dennis Riley, William D. McGhee, and Thomas Waldman

11. **Nucleophilic Aromatic Substitution for Hydrogen: New Halide-Free Routes for Production of Aromatic Amines** 133
    Michael K. Stern

12. **Chemistry and Catalysis: Keys to Environmentally Safer Processes** 144
    Leo E. Manzer

    TOOLS FOR ASSESSMENT OF BENIGN CHEMISTRY

13. **Alternative Syntheses and Other Source Reduction Opportunities for Premanufacture Notification Substances at the U.S. Environmental Protection Agency** 156
    Carol A. Farris, Harold E. Podall, and Paul T. Anastas

14. **Computer-Assisted Alternative Synthetic Design for Pollution Prevention at the U.S. Environmental Protection Agency** 166
    Paul T. Anastas, J. Dirk Nies, and Stephen C. DeVito

    INDEXES

Author Index 187

Affiliation Index 187

Subject Index 187

Preface

BENIGN BY DESIGN CHEMISTRY is synthetic chemistry designed to use and generate fewer hazardous substances. The ultimate goal of benign chemistry research is to develop and institute alternative syntheses for important industrial chemicals in order to prevent environmental pollution. With the United States chemical industry alone releasing more than three billion tons of toxic chemicals to the environment annually, significant challenges are available for chemists to design new syntheses that are less polluting. This type of prophylactic chemistry is as important to avoiding environmental problems as preventative medicine is to avoiding medical problems.

This book provides an opportunity for several chemists who are pioneers in the field of benign by design chemistry to present their basic research. We hope it will inspire many more chemists to become involved in creative environmentally responsible chemistry as it becomes the topic of more research and then moves further into industrial practice. Chapters in this book discuss the conceptual basis for benign chemistry and a wide range of research embracing this approach.

The symposium on which this book is based showcased results of several researchers who were the first to participate in a grant program founded by Roger L. Garrett of the Office of Pollution Prevention and Toxics at the U.S. Environmental Protection Agency. Garrett's vision allowed for the initiation and development of the Agency's program, Alternative Synthetic Design for Pollution Prevention. Other contributors highlight important results of benign chemistry research from academic, industrial, and governmental settings.

The symposium was the first one to explore this new area of research in synthetic chemistry. The numerous connections made between academic and industrial chemists at the symposium to establish research collaborations should hasten the further adoption of benign chemistry principles in industrial practice. The favorable audience reception, the extensive press coverage, the enthusiasm of our individual speakers, and the reaction of the scientific community following the symposium encouraged us to capture the essence of the symposium in this book. We acknowledge our authors and speakers for their cooperation and support in presenting the symposium and preparing chapters for this book.

This book should have broad appeal to chemists and others concerned with the manufacture and use of chemical products in an environmentally responsible manner. Certainly, basic researchers from academia, industry, and government should find the specific results and methodologies both useful and thought-provoking. The techniques and tools discussed in this book should also be of interest to those decision-makers who determine which syntheses will be used to manufacture chemical substances. In many ways, the message of the book—that dramatic advances can be made in pollution prevention by using the talents of synthetic chemists— is as important as the research itself.

Disclaimer

This book was edited by Paul T. Anastas and Carol A. Farris in their private capacities. No official support or endorsement of the U.S. Environmental Protection Agency is intended or should be inferred.

PAUL T. ANASTAS
CAROL A. FARRIS
Office of Pollution Prevention and Toxics
U.S. Environmental Protection Agency
Washington, DC 20460

September 13, 1994

In Memoriam

KENNETH G. HANCOCK was Director of the Division of Chemistry, Mathematics and Physical Sciences Directorate of the National Science Foundation (NSF). He died suddenly and unexpectedly on September 10, 1993, in Budapest, Hungary, while attending an international environmental chemistry workshop with colleagues from France, Hungary, and several Eastern European countries. Ken's death was a shock and loss to his family, his colleagues, and his friends. Ken, an excellent chemist, administrator, teacher, and dedicated civil servant, was a graduate *cum laude* from Harvard University in 1963. Following his graduate studies at the University of Wisconsin (Ph.D., 1968), he became a tenured associate professor of chemistry at the University of California—Davis. He joined the NSF staff in 1977. Ken played a leadership role in a number of areas at NSF, including international affairs and the fostering of environmentally benign chemical synthesis and processing.

Ken's commitment to green chemistry and pollution prevention has prompted the establishment of the Kenneth G. Hancock Pollution Prevention Fund to be administered by the American Chemical Society's Division of Environmental Chemistry, Inc. The Fund will "recognize student achievement in environmental chemistry, especially in environmentally benign chemical synthesis and processing" through scholarships to undergraduate and graduate students. Contributions may be made to the following address.

The Kenneth G. Hancock Pollution Prevention Fund
American Chemical Society
Development Office
1155 16th Street, N.W.
Washington, DC 20036

This book is one of several that are being published in his memory.

OVERVIEW

Chapter 1

Benign by Design Chemistry

Paul T. Anastas

U.S. Environmental Protection Agency, Office of Pollution Prevention and Toxics, Mail Code 7406, 401 M Street, Southwest, Washington, DC 20460

> The role of the synthetic chemist is crucial toward meeting the goals of both environmental protection and economic growth. The principles for the discovery, evaluation and development of environmentally benign alternative synthetic pathways have been investigated and applied both in basic research and in commercial practice. The principles of pollution prevention will be playing an increasing role in the routine work of the synthetic chemist in the future and each chemist will need to know how to design syntheses that are more environmentally benign in order to increase the chances of commercial viability of the methodology.

The synthetic chemist has generally not viewed him/herself as having a role in environmental chemistry. The message of Benign By Design chemistry is that the role of synthetic chemists is fundamental to environmental concerns and to pollution prevention. While the traditional areas of environmental chemistry, such as analytical and atmospheric chemistry, will always play an important role it is crucial that the synthetic chemists (who have been perceived as being responsible for much of the toxic pollution that exists) now be associated with the avoidance of environmental problems. It is essential from an environmental and economic standpoint that pollution prevention become the paradigm of first choice in the area of chemical production. Synthetic chemists are the only ones capable of instituting this fundamental change.

What is Pollution Prevention?

The United States's approach toward dealing with environmental problems has evolved since the early stages of the environmental movement in the 1960s and early 1970s. Most approaches have centered around the "command and control" approach to pollution. In its earliest form this involved the government allowing potential releasers of toxic substances to release materials only in certain limited amounts and/or requiring them to obtain permits to dispose of toxic chemicals, often to air or water. At the time, this tactic was described by the rather black-humor rhyme, "Dilution is the solution to pollution." As the environmental movement progressed, it became more common for the government to require treatment of various wastes prior to their release to the environment. Usually this involved sufficient treatment techniques such that the concentration of the toxic substance was reduced to an acceptable level. It is only within the last several years that the U.S. approach to dealing with pollution has been not to create the polluting substances in the first place. This is the basis of pollution prevention.

Pollution prevention as an approach to environmental problems is analogous to preventative medicine as an approach to medical problems. It is commonly accepted that it is preferable to avoid getting a disease rather than to have to cure the disease. It is equally logical and intuitive that avoiding environmental problems is far preferable to having to cure them. Both of these adages hold for many of the same reasons to both the patient and the environment: minimize pain and suffering, avoid long term systemic damage caused by the initial insult and reduce costs of sustained viability.

The costs of the command and control approach to environmental problems are staggering. Estimates of how much business is spending on control and treatment technologies are as high as $115 billion dollars annually (*1*). Still with all of this money invested in pollution control, over three billion pounds of waste were released to the environment in 1992, as shown in Table I, according to the U.S. EPA's Toxic Release Inventory (*2*), which has tracked the release of only approximately 300 chemicals. Since over 70,000 chemicals are currently in commerce in the United States (*1*), it is easily seen that despite efforts of regulatory agencies to control the release of chemicals to the environment, they are only capable of focussing on those few of highest priority. Most of these documented chemical releases have been to the air as shown in Figure 1.

It is certainly desirable for industry to reduce its operating costs associated with the compliance with local, state and federal regulations, waste treatment and waste disposal. By focussing on reducing the amount of waste that is generated, a company will be able to achieve economic benefits associated with avoiding these operating costs. This type of economic incentive is encouraging companies to look inward to find ways to regulate themselves and reduce their environmental releases. The private sector is finding that pollution prevention makes good business sense as evidenced by the examples of the Dow Chemical Corporation's WRAP Program (Waste Reduction Always Pays) (*3*) and 3M's 3P Program (Pollution Prevention Pays) (*4*). It is revolutionary in the fullest sense of the word when environmental stewardship is transformed from being perceived by industry as an economic burden to being perceived as necessary for increased profitability and competitiveness. This is the fundamental difference between pollution prevention and the previous command and control approaches to dealing with environmental problems and this is why there is the promise of profound effectiveness with this approach.

Table I. TRI Chemical Releases

1992 Releases	Pounds
Total Releases	3,181,646,757
Air Emissions	1,844,958,336
Surface Water Discharges	272,932,953
Underground Injection	725,946,415
Releases to Land	337,809,053

SOURCE: U.S. EPA, TRI Annual Report, 1992

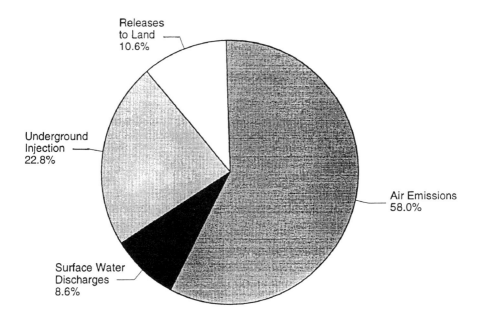

Figure 1. Releases of TRI Chemicals to Various Media
SOURCE: U.S. EPA, TRI Annual Report, 1992

In 1990 the U.S. Congress passed legislation to promote pollution prevention as the basis for environmental policy in the United States. In the Pollution Prevention Act of 1990 (5), pollution prevention is defined in terms of source reduction. Source reduction is defined as:

> Any practice which reduces the amount of any hazardous substance, pollutant, or contaminant entering any waste stream or otherwise released into the environment (including fugitive emissions) prior to recycling, treatment, or disposal; and reduces the hazards to public health and the environment associated with the release of such substances, pollutants or contaminants. The term includes equipment or technology modifications, processes or procedure modifications, reformulation or redesign of products, substitution of raw materials and improvements in housekeeping, maintenance, training, or inventory control.

It was through this statute the U.S. EPA was first mandated to pursue pollution prevention solutions in all of its environmental protection initiatives. The Administrator of the EPA, Carol Browner has cited pollution prevention as EPA's "central ethic" and has encouraged its incorporation into all future environmental regulations (6).

Some of the earliest environmental laws to incorporate pollution prevention were passed by state legislatures. States such as Massachusetts and New Jersey promulgated laws which instituted provisions for "toxic use reduction" in efforts to minimize the degree to which hazardous substances are employed at various stages of manufacturing, processing and use of chemical products (7).

Early Approaches to Pollution Prevention

When the concept of pollution prevention first was introduced as an approach to environmental problems, the majority of its early manifestations were in the form of housekeeping solutions. Reducing leakages in piping systems, covering vats and vessels which hold volatile substances to reduce evaporation, reducing loss of material through over-spray in spraying applications were some of the earliest pollution prevention practices. Many of these process changes resulted in significant reductions in waste and pollution at the source. Materials which otherwise would have been treated or disposed of could now be used profitably. While solutions such as these seem in retrospect simple common sense, they were none-the-less, changes in standard operating procedures for many companies.

Many of these early approaches to pollution prevention are what could be described as the "low-hanging fruit:" in other words, those solutions which are fast and easy to implement. The potential for pollution prevention, however, is far more fundamental. It requires a change in the manner in which products and processes are designed from their inception. Without question, all approaches toward preventing pollution should be assessed and implemented wherever possible. This chapter, however, will focus on the earliest design phase of chemical manufacture, the design of the synthetic sequence.

The Future of Pollution Prevention

As the familiarity with the concept of pollution prevention increases throughout the scientific and industrial community, source reduction is being considered at earlier and earlier stages of the life cycle. Products and processes are going to need to be designed such that they do not generate waste in the first place. Even the lack of waste generation alone is not enough; products must be designed such that they do not use, either in their manufacture or use, hazardous substances. By eliminating a demand for the use of substances of this type there will be a multiplying affect on the environmental benefit of these new design procedures.

These choices must be made after careful analysis of the trade-offs that need to be made in the decision of how to synthesize a new chemical substance. While it is easy to state, correctly, that it is imperative to minimize the use and generation of substances which pose a hazard to human health or the environment, only those individuals qualified to fully understand the nature of the choices can be relied on to make those choices responsibly. This is precisely why the synthetic chemist will play an increasingly important role in allowing the chemical industry to discover and commercialize technical innovations. These innovations will need not only to maintain and improve on the quality of current products but also to develop new synthetic methods for these products to be made in a less costly and environmentally responsible manner. These principles will need to be built into the development protocols of new chemical products as well.

How Does Pollution Prevention Relate to the Synthetic Chemist?

To fully appreciate the fundamental role of the chemist in pollution prevention, we need to gain historical perspective and ask, "How have synthetic chemists traditionally designed and evaluated chemical syntheses for the manufacture of chemical products?" Over the years, chemists have repeatedly demonstrated their expertise to identify, understand and solve problems. In everything from pharmaceuticals to plastics, chemists have developed new methods and materials which have advanced society in countless ways. What is beginning to be recognized within the scientific community is that synthetic chemists will need to focus attention on ways of preventing environmental problems as effectively as chemists in general have been at solving them (8-17).

The Importance of Yield. One of the primary criteria for evaluating a synthetic transformation or an entire synthetic pathway in the manufacture of a chemical product is the yield of the process. Yield, simply stated, is the percentage of product obtained versus the theoretical amount one could have obtained for a given amount of starting material. This evaluation tool has been used historically because it is sound from a scientific as well as an economic perspective. From a scientific point of view, yield can be a good indicator of thermodynamic favorability of a particular process when evaluated in the context of the reaction or manufacturing conditions. From an economic point of view, yield is, of course, important as an indicator of the efficiency of use of the feedstocks. If the yield is low, other economic or technological factors must be considered or alternatives pursued to ensure that the inherent inefficiency doesn't result in an economically disadvantageous situation for the manufacturer.

The Importance of Feedstock Cost. Historically, the decision about which way to make a particular chemical substance would hinge on the selection of a feedstock or feedstocks. If there were a feedstock that was readily available and inexpensive, the synthetic methodology employed in the manufacturing process would usually reflect that economic logic. There is no doubt that the selection of feedstocks based on cost and availability and the use of yield in evaluating a synthetic scheme are always going to play an important role in the world of chemical manufacture. It is now the case, however, that these will no longer be the only criteria by which to judge a synthetic method.

How Has the World of Chemical Manufacturing Changed?

Since the time immediately following World War II when the chemical industry in the United States began to emerge as a dominant industrial sector, there have been significant changes in the societal values and the business climate of the United States. The same industry, the chemical industry, that was once hailed as the provider of modern convenience and innovation, now is associated with fouling the planet. To some degree these sentiments are justified by examining which industries are releasing the majority of toxic substances to the environment as shown in Figure 2. With these social opinions in place, laws soon followed which reflected the popular attitudes. The results of these statutes have effectively required a chemical company to consider additional factors when designing, producing and selling a chemical product. These considerations are listed in Table II.

Table II. Considerations of Costs of Chemical Manufacture

OLD	NEW
Feedstock Price/Availability	Feedstock Price/Availability
Energy Costs	Energy costs
	Regulatory compliance costs
	Waste disposal costs
	Waste treatment costs
	Liability costs
	Green marketing
	Consumer backlash

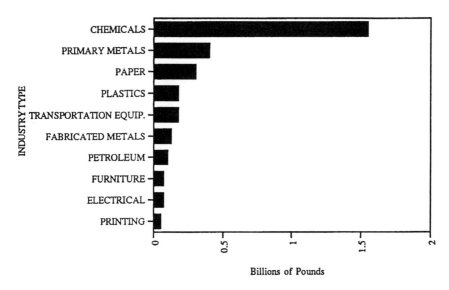

Figure 2. Top Ten Industries for Total TRI Releases, 1992
SOURCE: U.S. EPA, TRI Annual Report, 1992

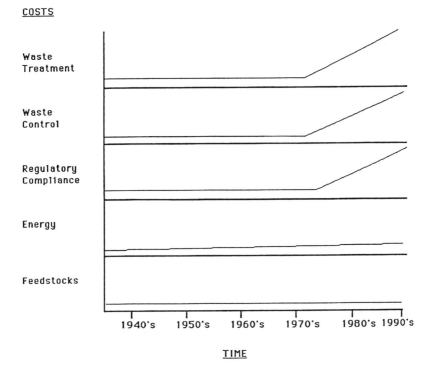

Figure 3. The Changing Costs of Chemical Manufacturing

Why is the Role of the Chemist Fundamental to Pollution Prevention?

In moving toward a society that is geared toward instituting pollution prevention principles on a national level, it is imperative that there be a consideration of how chemicals are made. This includes a consideration of the individual synthesis or the entire synthetic sequence involved in the synthesis of a chemical substance. Since the beginnings of the scientific approaches to chemistry, chemical synthesis steps and their mechanisms have been the primary focus of chemists. Historically, and into present times, the most important criterion for selection of one synthetic step in the formation of a chemical versus an alternative in small scale synthesis has been which one had the better "yield." (In the case of industrial synthesis both yield and cost of starting materials had been important determinants as mentioned above.) Using this criterion, thousands of synthetic chemical reactions have been studied and reported in the scientific literature. This collection of synthetic transformations is what the synthetic chemist uses when designing how to make a particular compound or how to add a particular functional group onto a compound. From a scientific standpoint, the use of yield as a criterion was kinetically and thermodynamically sound and in most instances it was favorable from an economic standpoint as well.

In view of the new emphasis on pollution prevention both by regulatory agencies and the chemical industry and the skyrocketing costs of waste disposal, waste treatment, and regulatory compliance, the evaluation scheme of judging the choice of a particular synthetic method based on the concept of maximum yield as the sole driving economic force is no longer valid. It is beginning to be recognized that it is possible to change the way that chemicals are made by changing the manner in which individual synthetic transformations or overall synthetic schemes are selected to make a chemical compound. These new methodologies can be designed so that they are intrinsically more environmentally benign. In large part, many of the basic synthetic methodologies are developed in academic laboratories and in those chemical companies large enough to have significant chemistry research and development departments. Pursuing this area of benign chemistry research will provide the chemist with more "tools" or synthetic methodologies to select from when integrating pollution prevention into a manufacturing process.

This area of research meets both the chemical industry's and society's needs in developing the concept of pollution prevention and the academic community's need to focus on basic research. The next generation of synthetic chemists will certainly be focused on how to build new chemical structures but they will also be incorporating all of the impacts, scientific, economic and environmental, into their selection of how to make the chemical product.

The moment that a chemist puts pencil to paper to design how a chemical product will be made, he/she is intrinsically making decisions about:

- What hazardous wastes will be generated,
- What toxic substances will need to be handled by the workers making the product,
- What toxic contaminants might be in the product,
- What regulatory compliance issues there are in making this product,

- What liability concerns there are from the manufacture of this product, and
- What waste treatment costs will be incurred.

By putting forethought into the selection of the method of making a chemical product such that all of the scientific, environmental, and economic impacts of a particular process are considered, the synthetic chemist can have perhaps the most influence in achieving pollution prevention.

What is Benign By Design Chemistry?

Benign By Design chemistry defines synthetic elegance on the basis of three factors:
- **Efficiency** of synthetic methodology
- **Economically** viable
- **Environmentally** Benign

The concept of synthetic efficiency has been discussed in terms such as carbon economy and atom economy (*18*). Both of these concepts poignantly illustrate the desirability of the incorporation of all of the atoms used in the transformation of the starting materials into the product. While full incorporation of atoms will never be achieved in all synthetic methodologies, it should be used as a goal and as one criterion by which to judge the benign nature of a reaction or reaction scheme.

Economic viability, simply stated is a pass/fail test for commercialization of a process to manufacture chemical products. If the synthetic technology cannot survive economically, the other virtues of the method quickly become irrelevant. It must be remembered, however, that the economic analysis must make sure to take into consideration all costs related to the manufacture such as those listed above (e.g., waste disposal, regulatory compliance, etc.). Without factoring in these significant related costs, it would be easy to incorrectly dismiss a new, more environmentally benign methodology as not being cost competitive (Figure 3).

What are the new considerations in designing a new synthetic pathway for the manufacture of a chemical product? As has been discussed above, one can no longer consider just the yield and cost of feedstocks when selecting a synthetic route to the manufacture of a chemical product. Synthetic chemists and decision makers need to ask the following questions:

What are the Toxicity Impacts of the Manufacture to Humans? This analysis must include all substances related to the synthesis. Toxic endpoints should include not only lethality (LD_{50}), but also endpoints such as neurological disorders, reproductive and developmental effects, etc.

What is the Impact on the Living Environment? Considerations of direct toxicity to various plant life and wildlife should be included whenever possible.

What is the Impact on the Larger Environment? The "larger environment" would include the effects such as stratospheric ozone depletion, atmospheric ozone generation, greenhouse gas generation, acidification and/or deoxygenation of lakes and streams, etc.

Will This Increase the Potential for a Chemical Accident? While very often pollution prevention and accident prevention work hand-in-hand to minimize the risk to human health and the environment, this is not always the case and cannot be assumed without an analysis.

What Aspects of the Synthetic Scheme Need to be Evaluated for Whether It Is Environmentally Benign?

For many years, much of the focus of the environmental movement and specifically the regulatory agencies was concentrated on the chemical products that the chemical manufacturers were producing to ensure that they did not pose a risk. Now the focus is no longer just the final product but all of the substances associated with the manufacturing process. A few of the materials to be considered are listed below.

- Feedstocks
- Reagents
- Reaction Media
- By-products and Impurities
- Catalysts
- Separation Solvents
- Distillation Products

Approaches to Benign By Design Chemistry

Academic Research. There has been a significant amount of activity in the area of Benign By Design chemistry in recent years. While it is definitely an area of investigation in its infancy, it is also being recognized as providing rich opportunities for new basic research in the academic community. Some of this research is being developed and pursued with the goals of pollution prevention and designing environmentally benign synthesis in mind. Other research projects have been pursued in years past and are now being recognized as having significant technological advantages which make them more environmentally benign than alternative techniques.

An example of the latter is the work in stereoselective reaction schemes, including chiral catalysis. This has been an active area of research for decades for a variety of good reasons. An additional good reason is that by making only the enantiomer that is desired from an otherwise racemic mixture, one is preventing the wasteful production of an equal amount of useless product. Work in general catalysis, biomimicry, and solid state chemistry has been pursued previously with a variety of goals and objectives to which now can be added environmental benefits.

Several areas of research have been more thoroughly investigated for their application in Benign By Design Chemistry than others.

Computer Design of Synthetic Methodologies. Over the past twenty-five years, the concept of developing and using computer software to design synthetic transformations or entire synthetic pathways has been attempted by a number of different research groups *(19-22)*. Some of these computer programs work through use of databases, some by heuristic logic programs and some through the use of artificial intelligence. These programs may be very specific to certain reaction types or may be very general and may work in the synthetic or retro-synthetic direction. Until recently, however, there has not been a consideration of environmental impacts in the evaluation schemes to choose one synthetic transformation over another. The U.S. EPA has embarked on a project to promote the incorporation of environmental considerations into the major software programs as well as the development of new software to design environmentally benign syntheses *(23)*.

Solvent Alternatives. The environmental consequence of using organic solvents in the manufacture of chemical products has been an issue of concern for many years. The new Clean Air Act Amendments (CAAA) *(24)* have listed many commonly used volatile organic compounds (VOCs) which are used as solvents as hazardous air pollutants. A number of research projects are on-going with the goal of reducing the amount of VOCs used and released by the chemical industry.

One active area is in using super-critical fluids (SCFs) as a reaction medium. While the usefulness of SCFs as an extraction solvent, a cleaning solvent or in analytical methodologies has been well-established, the use of super-critical carbon dioxide as well as other SCF's is a far less explored area of research. There have been recent successes documented in the use of SCFs as a reaction medium for polymerization reactions *(25)*, free-radical transformations *(26)*, and in certain catalytic transformations *(27)*.

One approach toward solvent alternatives to VOCs is the increased use of aqueous reaction systems. There have been a number of investigations on conducting synthetic transformations in water which have previously only been carried out routinely in organic solvents *(28)*.

With the goal of VOC solvent reduction outlined by both the environmental movement and the regulatory community, an obvious approach to dealing with the problem is through the greater use of solventless and solid state chemistry. Neat reactions have the obvious advantage of not generating any waste solvent that needs to be incinerated, disposed of or recycled while solid-state chemistry has the additional advantage of having a very low vapor pressure. The low vapor pressure decreases the exposure of workers to any hazard that may exist through inhalation, thereby reducing the risks overall.

Alternative Feedstocks. One area to address when evaluating a synthetic transformation or a synthetic pathway to assess whether or not it is environmentally benign is what materials are being employed at the front-end, the feedstocks. By using substances which either reduce or do not possess significant hazards to human health or the environment, the synthetic chemist is reducing the overall risk of the manufacturing process. There are several areas that are prime areas for research on benign feedstocks.

For a number of years, there has been research in the area of using "masked synthons" in synthetic schemes to achieve difficult transformations. These synthons could be designed to withstand reaction conditions otherwise unfavorable to the parent functional group but still serve as the synthetic equivalent in order to accomplish the synthesis. Using this same logic, it is possible for a masked synthon to be used in order to make the chemical, which must be handled by individuals involved in the manufacturing process, less toxic. Small structural changes in a feedstock can reduce the toxicity of the substance by orders of magnitude while still allowing the synthetic equivalent to be generated in situ.

The vast majority of the chemical products produced in the United States today are ultimately derived from petroleum feedstocks. While there are petroleum feedstocks that should be recognized as being environmentally benign, there are many others that are quite hazardous such as benzene, a known carcinogen. It is for this reason that there is research being conducted on alternative feedstocks to petroleum such as biological starting materials (29-30). Examples of the potential applications of biological feedstocks are shown in Figure 4. Biological feedstocks provide several advantages including the fact that they are derived from renewable sources. In addition, most petroleum products are in a highly reduced state and need to be oxidized in order to instill the type of functionality necessary for further product development. Oxidation processes can have pronounced environmental impacts, especially those which employ the use of heavy metals. In contrast, biological feedstocks are often highly oxidized and highly functionalized which, generally, allows for cleaner types of transformations such as reductions.

Alternative Catalysis. Catalysis has the promise of making an increasing number of chemical manufacturing processes not only more efficient, but also more environmentally benign. There has been investigation into new applications of non-traditional catalysts to reduce the environmental impacts of certain reaction types. The Friedel-Crafts reaction is a widely used synthetic methodology which classically incorporates the use of a Lewis Acid catalyst as a part of its mechanism. Kraus (31) has carried out research in the area of producing "Friedel-Crafts-type" acylation products using light as a catalyst. Other work on the cleavage of dithianes has been performed by Epling's group (32). Their methodology changes the typical "reductive cleavage" methodology which uses heavy metals as catalysts and uses visible light with a dye in order to affect the cleavage.

Industrial Initiatives

Catalysis. One important area of investigation that is being vigorously pursued by the chemical industry is catalysis. Certainly, catalytic processes can offer many advantages, one of which is making the process more environmentally benign in many cases. The work being carried out by Catalytica (33) and Chemical Contractor Ltd. (34) in designing new catalysts which have the net effect of decreasing the environmental impact of a chemical process is illustrative of the work that is being done.

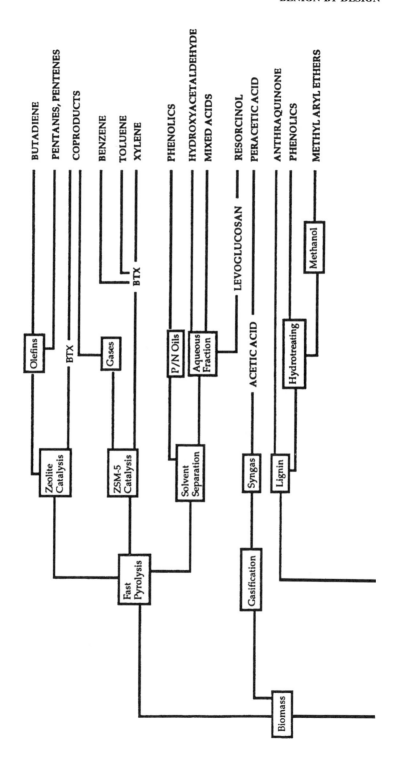

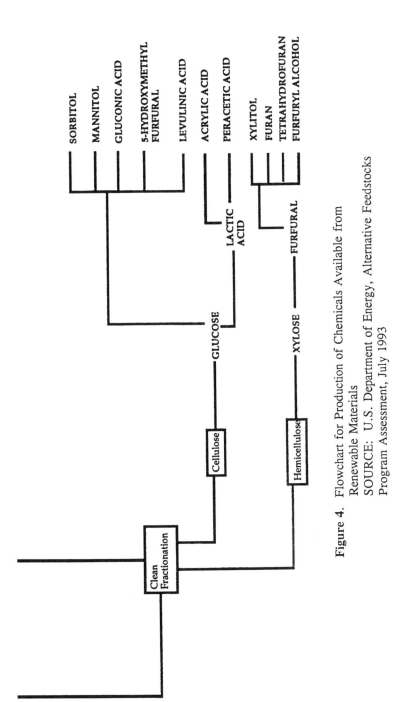

Figure 4. Flowchart for Production of Chemicals Available from Renewable Materials
SOURCE: U.S. Department of Energy, Alternative Feedstocks Program Assessment, July 1993

Alternative Feedstocks and Processes. The chemical industry is beginning generally to review the feedstocks and processes that they have historically used in the manufacture of some of their most basic products. Monsanto has begun to institute changes in the way the company makes aromatic amines (35,36). Rather than proceed through a route that necessarily generates and utilizes chlorinated aromatics, many of which have been identified as posing environmental hazards, Monsanto is developing a method of direct amination of nitrobenzene via nucleophilic aromatic substitution of hydrogen. In addition, Monsanto is also pursuing methods of making urethanes and isocyanates without the use of the acutely toxic phosgene by replacing it with carbon dioxide (37).

DuPont is using Just-In-Time manufacturing methods which incorporate in situ generation techniques to minimize the risk of exposure to hazardous substances (14). These techniques have been applied to processes which use particularly toxic substances such as methylisocyanate in an agrochemical process. This example illustrates efforts being pursued in basic research both in academia (described above) and in industry in developing environmentally benign masked synthons.

Some widely used materials in the chemical industry are also commonly recognized as posing significant hazard. Fluorinating agents such as HF, $FClO_3$ and CF_3OF are well known for both their efficacy and their hazards. Air Products has developed a fluorinating reagent which requires no special handling and performs selective fluorinations on a wide variety of substances. Selectfluor (38), or 1-chloromethyl-4-fluoro-1,4-diazonia[2.2.2]bicyclooctane bis(tetrafluoroborate), is an example of product development which includes both effectiveness and environmental considerations.

Governmental Initiatives and Public/Private Partnerships

The goals of pollution prevention complement many of the strategic goals of the United States. It is obvious that the national goals of environmental protection as reiterated in the Pollution Prevention Act are served by this relatively new focus of the environmental movement. However, it is also true that interests of the country's economic competitiveness, health and well-being of the nation's populace as well as its basic science and research capabilities are also beneficiaries of this new pollution prevention policy. It is for these reasons that there is such a widespread support for more environmentally benign processes to be developed throughout all sectors of the economy and the scientific community.

U.S. Environmental Protection Agency. In the earliest days of the Pollution Prevention movement, the U.S. EPA's Office of Pollution Prevention and Toxics began the pursuit of the concept of using alternative synthetic pathways for pollution prevention. The EPA structured a model funding program which would promote Benign By Design chemistry in the development of new synthetic methodologies. This program was designed to serve as demonstration to both the scientific community as well as the major scientific research funding agencies, that this type of research was viable, productive and necessary. The initial six grants provided ample justification for further funding and served as a vivid illustration of the type of fundamental research that can be accomplished in this area. This initial effort was the forerunner to the large number of programs which are pursuing the goal of designing chemical syntheses to achieve pollution prevention.

Benign Synthetic Design Tool Development. The EPA is engaged in the development of a number of tools in the form of guidance, computer software, and evaluation protocols to assist the synthetic chemist in designing benign reactions. As discussed previously the EPA is promoting the incorporation of environmental considerations into synthetic software programs such that when various synthetic pathways are generated, a listing of environmental impacts is presented to the chemist to consider.

Guidance and Evaluation Protocols. As part of the EPA's Office of Pollution Prevention and Toxics New Chemicals Program there has been developed a procedure which reviews new chemical substances entering United States commerce for "unreasonable risk to human health and the environment." As part of this program, a Benign By Design Chemistry review has been instituted to review the manufacturing process. This review, Synthetic Methodology Assessment for Reduction Techniques (SMART) (*39*) provides suggestions to the chemical industry for ways to incorporate more benign methods of making, processing or using chemical substances. The review procedure is to be converted into a guidance manual such that the regulated community can use the protocols in evaluating both their new and existing manufacturing processes.

Curriculum Development. The skill of incorporating environmentally benign synthetic techniques into the manufacturing processes of the chemical industry will be required well into the foreseeable future. In order to make these pollution prevention approaches a systematic part of doing business, industry is going to need to be equipped with a workforce that has been trained in Benign By Design Chemistry. For these reasons, the EPA has promoted the development of educational materials which will train chemists at various levels of their education in considering environmental impacts. The materials will include:

- Textbook supplements which parallel the classical chemistry texts but offer environmentally benign alternatives to standard techniques;

- A reference module for faculty which allows them to easily translate late-breaking environmentally benign research into their classroom presentations;

- Laboratory modules which illustrate the experimental principles of benign chemical synthesis through undergraduate lab experiments; and,

- Professional training for industrial bench chemists to address why pollution prevention is desirable for a company and how environmentally benign chemistry can achieve it.

Environmental Technology Initiative. The current Administration launched a program to promote the development of innovative technologies which address environmental problems. The Environmental Technology Initiative (ETI) was begun in 1993 and designed to provide funding which will facilitate the generation and utilization of environmental technologies through research, development, information dissemination and removing regulatory barriers. This program includes, as one aspect, the development of more environmentally benign synthetic chemistry methodologies. The EPA acts as a steward for the funding and sets criteria by which funding can be distributed to other Federal or State agencies as well as public/private consortia.

National Science Foundation. As a major funding institution for basic research in chemistry including chemical synthesis, the National Science Foundation (NSF) has identified environmental considerations in designing synthetic methodologies as an area that warrants support and funding. Through the use of a special program for funding developed in the Division of Chemistry by Dr. Kenneth A. Hancock entitled "Environmentally Benign Chemical Synthesis and Processing Program," the NSF is seeking to promote the use of environmentally friendly methods in chemical manufacturing through the discovery of innovative chemical technology. The NSF developed this program in concert with the Council for Chemical Research (CCR) and established a Memorandum of Understanding in January of 1993 with the EPA to work collaboratively in promoting benign chemistry.

The NSF also has promoted benign chemical manufacturing through the use of its Industry/University Cooperative Research Centers Program. An example of one of these centers would be the Emission Reduction Research Center housed at the New Jersey Institute of Technology (NJIT). This collaboration between academic institutions such as NJIT, Massachusetts Institute of Technology, Ohio State University and Pennsylvania State University with several specialty chemical and pharmaceutical companies and Federal agencies such as NSF and EPA provides a useful venue for conducting research on environmentally benign manufacturing methods and ensuring that the research fits the needs of industry and society.

Department of Energy. As part of its historical charge toward exploring all aspects of energy efficiency and general energy utilization, the Department of Energy (DOE) has focussed much attention on the environmental impacts of various technologies. In its program, Environmentally Conscious Manufacturing, DOE works with a wide range of companies from many industrial sectors. In addition, the National Laboratories are actively pursuing research in environmentally benign chemistry. As an example, Los Alamos National Laboratory has established a partnership with the U.S. EPA's Office of Pollution Prevention and Toxics in investigating the use of supercritical fluids as a synthetic reaction medium. This collaborative effort includes private companies as well as universities and serves to illustrate an approach to conducting fundamental research which can be geared to both the need for technological innovation and environmental responsibility.

Future of Benign By Design Chemistry

A workshop was convened jointly by the U.S. EPA and the NSF entitled "Green Syntheses and Processing in Chemical Manufacturing" (40). Here, experts in the field of chemical synthesis from academia, industry and government were asked to identify the needs, in terms of research and implementation, which have to be addressed in order to accomplish the incorporation of environmentally benign chemistry in the chemical manufacturing industry. Many of the statements of need are very specific while others are quite global in their scope. Some of the suggestions identify work currently being done which needs to be identified and applied to environmentally benign applications. The list below should not be viewed as comprehensive by any means, but rather as a thought-provoking starting point.

- Identify a list of the top processes that provide opportunities for pollution prevention;

- Include benign chemistry as a core tenet of the chemistry curriculum; make responsible chemistry as important as creative chemistry.

- Develop robust homogeneous oxidants;

- Develop safer solvents to replace current ones - or use solventless systems;

- Use selective clean oxidative functionalization (bond making processes) in order to reduce the use of substitution processes;

- Develop and promote the use of solid acid catalysts (heterogenizing processes);

- Design processes for the recovery of reagents - develop new separation techniques;

- Investigate novel media and/or reaction systems to enhance selectivity of chemical transformations;

- Develop educational approaches to familiarize synthetic chemists with separations and processes (physical/chemical isolation);

- Develop an valuative tool to assess environmental impact of synthetic methods on both the bench and process scale;

- Develop a means of funding interdisciplinary teams of scientists to collaborate on addressing key problems in research and education;

- Develop a database/information network on benign synthesis;

- Conduct research on small molecule biomimetic catalysts;

- Develop catalysts capable of complex reactions by multi-step transformations at a single site;

- Develop functional mimics of enzymes;

- Investigate means to increase enantioselective synthetic methods (particularly catalysis);

- Promote investigations in biocatalysis, enzymatic, and microbiological transformations;

- Promote photochemistry and electrochemistry for benign synthesis;

- Investigate the use of supercritical fluids in catalytic and biocatalytic reactions;

- Investigate molecular design for advanced separations;

- Design target molecules which preserve the desired function while mitigating toxicity by structural as well as physical/chemical property modification;

- Develop, among synthetic chemists, a functional group understanding of health and environmental hazards; and

- Investigate the use of renewable feedstock alternatives to petroleum.

Conclusion

Synthetic chemists have an important and fundamental role to play in the environmental movement. Because of their training, knowledge, and expertise, they are the only people capable of designing chemical syntheses at the front-end to ensure that environmental impacts are minimized. As mentioned previously, Benign By Design Chemistry is not a panacea that will solve the world's environmental problems. There are numerous situations where other pollution prevention solutions or even pollution control measures may need to be employed due to cost or logistical considerations. However, Benign By Design Chemistry should be the option of first choice which is built into the earliest stages of planning to manufacture a chemical product in order to ensure full consideration of the most fundamental pollution prevention methods available.

Acknowledgments

I would like to express my sincere gratitude and appreciation to Dr. Roger L. Garrett for his foresight and vision in establishing the Alternative Synthetic Pathways for Pollution Prevention Project and for putting his faith in a young chemist to develop the program.

Disclaimer

This chapter was prepared by Paul Anastas in his private capacity. No official support or endorsement of the U.S. Environmental Protection Agency is intended or should be inferred.

Literature Cited

1. Underwood, J.D. *EPA Journal* **1993**, *19(3)*, pp 9-13.
2. U.S. EPA. *1992 Toxics Release Inventory: Public Data Release*; EPA 745-R-94-001; U.S. EPA: Washington, DC, 1992.
3. Dow Chemical Company. *Environmental Health & Safety Milestones: Dow's History of Commitment.* **undated**, Cited In U.S. EPA. *Pollution Prevention 1991: Progress on Reducing Industrial Pollutants*; U.S. EPA: Washington, DC, 1991, EPA 21P-3003; p 58.
4. 3M. *3M's Pollution Prevention Pays*. Environmental Engineering and Pollution Control Dept./3M: St. Paul, MN, undated.
5. Pollution Prevention Act of 1990. 42 U.S.C. §§13101-13109, **1990**.
6. Browner, C.M. *EPA Journal* **1993**, *19(3)*, pp 6-8.
7. U.S. EPA. *Pollution Prevention 1991: Progress on Reducing Industrial Pollutants*; EPA 21P-3003; U.S. EPA: Washington, DC, 1991, pp 83-84.
8. Amato,I. *Science (Washington, D.C.)* **1993**, *259*, pp 1538-1541.
9. Illman, D.L. *Chem. Eng. News* **1993** (March 29), *71(13)*, pp 5-6.
10. Ember, L. *Chem. Eng. News* **1991** (July 8), *68(28)*, pp 7-16.
11. Ember, L. *CHEMTECH* **1993** (June), *23(6)*, p 3.
12. Wedin, R. *Today's Chemist at Work* **1993** (Jul/Aug), pp 16-19.
13. *Dallas Morning News* **1993** (Sept. 13), p 1D.
14. Illman, D.L. *Chem. Eng. News* **1993** (Sept. 6), pp 26-30.
15. Woods, M. *Sacramento Bee* **1993** (August 26), p A18.
16. *Business Week* **1993** (August 30), *3334*, p 65.
17. Rotman, D. *Chem. Week* **1993** (Sept. 22), *153(10)*, pp 56-57.
18. Trost, B.M. *Science (Washington, D.C.)* **1991**,*254*, pp 1471-1477.
19. Hendrickson, J.B. *J. Am. Chem. Soc.* **1986**, *108*, pp 274-281.
20. Hendrickson, J.B.; Miller, T.M. *J. Am. Chem. Soc.* **1991**, *113*, pp 902-910.
21. Jorgensen, W.L.; Laird, E.R.; Gushurst, A.J.; Fleicher, J.M.; Gothe, S.A.; Helso, H.E.; Padreres, G.D.; Sinclair, S. *Pure Appl. Chem.* **1990**, *62*, pp 1921-1932.
22. Corey, E.J.; Long, A.K.; Rubenstein, S.D. *J. Org. Chem.* **1985**, *228*, pp 408-418.
23. See chapter in this book by DeVito, S.C.; Nies, J.D.; Anastas, P.T.
24. Clean Air Act Amendments of 1990 (P.L. 101-549). *42 U.S.C. §7401, et seq.*, **1990**.

25. DeSimone, J.M.; Maury, E.E.; Menceloglu, Y.Z.; McClain, J.B.; Romack, T.J.; Combes, J.R. *Science (Washington, D.C.)* **1994**, *265*, pp 356-361.
26. Tanko, J.M.; Blackert, J.F. In *Preprints of Papers Presented at the 206th ACS National Meeting, Chicago, IL, August 22-27, 1993*; Elzerman, A.W., Chairman; American Chemical Society: Division of Environmental Chemistry: Milwaukee, WI, 33(2); pp 313-15.
27. Tumas, W.; Feng, S.; LeLacheur, R.; Morgenstern, D.; Williams, P.; Buelow, S.; Burns, C.; Foy, B.; Mitchell, M.; Burk, M.; Waymouth, R.; In *Preprints of Papers Presented at the 208th ACS National Meeting, Washington, DC, August 21-25, 1994*; Bellen, G.E., Chairman; American Chemical Society: Division of Environmental Chemistry: Milwaukee, WI, 34(2); pp 211.
29. Draths, K.M., Ward, T.L.; Frost, J.W. *J. Am. Chem. Soc.* **1992**, *114*, pp 9725-9726.
30. *Alternative Feedstocks Program Technical and Economic Assessment*; Bozell, J.J.; Landucci, R., Eds.; U.S. Department of Energy: Boulder, CO, 1993; pp 211-217.
31. Kraus, G.A.; Kiriyuki, M.; Wu, Y. In *Preprints of Papers Presented at the 206th ACS National Meeting, Chicago, IL, August 22-27, 1993*; Elzerman, A.W., Chairman; American Chemical Society: Division of Environmental Chemistry: Milwaukee, WI, 33(2); pp 330-331.
32. Epling, G.A.; Wang, Q. In *Preprints of Papers Presented at the 206th ACS National Meeting, Chicago, IL, August 22-27, 1993*; Elzerman, A.W., Chairman; American Chemical Society: Division of Environmental Chemistry: Milwaukee, WI, 33(2); pp 328-329.
33. Cusamano, J. *CHEMTECH* **1992**, *22*, pp 482-489.
34. *Chemicaoggi* **1992** (Mar.), pp 44-45.
35. Stern, M.K.; Bashkin, J.K.; U.S. Patent 5117063A, **1992**, to Monsanto.
36. Stern, M.K.; Hileman, F.D.; Bashkin, J.K. *J. Am. Chem. Soc.* **1992**, *114*, pp 9237-9238.
37. McGhee, W.D.; Riley, D.P. *Organometallics* **1992**, *11*, pp 900-907.
38. Lal, G.S. *J. Org. Chem.* **1993**, *58(10)*, pp 2791-6.
39. See chapter in this book by Farris, C.A.; Podall, H.E.; Anastas, P.T.
40. U. S. EPA. *Proceedings: Workshop on Green Syntheses and Processing in Chemical Manufacturing, Cincinnati, OH, July 12-13, 1994*; U.S. EPA: Cincinnati, OH, **1994,** in press.

RECEIVED September 13, 1994

Chapter 2

Environmentally Benign Chemical Synthesis and Processing for the Economy and the Environment

Kenneth G. Hancock[1] and Margaret A. Cavanaugh

National Science Foundation, 4201 Wilson Boulevard, Arlington, VA 22230

A new era of university-industry-government partnerships is required to address the intertwined problems of industrial economic competitiveness and environmental quality. Chemicals that go up the stacks and down the drains are simultaneously a serious detriment to the environment, a waste of natural resources, and a threat to industrial profitability. Recently, the NSF Divisions of Chemistry and Chemical and Transport Systems have joined with the Council for Chemical Research in a grant program to reduce pollution at the source by underwriting research aimed at environmentally benign chemical synthesis and processing. Part of a broader NSF initiative on environmental science research, this new program serves as a model for university-industry-government joint action and technology transfer. Other features of this program and related activities are discussed.

Chemistry is in the news, and the news isn't always good. Though today's high standard of living rests firmly on the creative contributions of chemists -- from food preservatives to pharmaceuticals, from crop-enhancing agrochemicals to synthetic fibers and plastics -- chemists have also created, as byproducts of their prodigious productivity, a host of environmental problems. Some of the most publicized in recent years have been the depletion of stratospheric ozone by CFCs, global warming by greenhouse gases, chemical spills in the Rhine River, and nitrous oxide as a byproduct of nylon production.

With continuing innovation, several hundred new chemicals are introduced each year, while thousands of new stacks and pipes release chemical effluents into the air, soil, and water. These new products, like their predecessors, are the building blocks of a technology-based economy. To satisfy global demand, the U.S. chemical industry employs nearly 2 million (or 11%) of all U.S. manufacturing workers and generates a $15 billion positive trade balance annually (*1*). Clearly, the economic vitality of the chemical industry must be maintained, and at the same time the behavior of chemicals in

[1]Deceased

the environment must be determined in order to avoid the risks they might pose to humans and other organisms (2).

In addition to the many sound health and economic reasons to worry about the environment, there is another urgent moral imperative: we have a responsibility to preserve the environment and a fair share of its natural, non-renewable resources for our children and for the generations that follow.

Social and Economic Pressures

The magnitude of environmental problems -- around the world -- is daunting. Increasingly, the public is aware of these problems, worried about the future, and ready to demand action, not only from politicians but also from the science and technology community. Some problems are within our present scientific and engineering capabilities if we have the public will to act. Others are not. The need for science and engineering research related to the environment is not only great, but urgent. Moreover, the U.S. industrial base, built in many cases around older technology than the industries of our newer manufacturing competitors, needs new knowledge and technology to remain economically competitive.

Some environmental problems, many of the most severe, transcend national boundaries, so their solution will require international cooperation on a scale rarely achieved. Because the time span from pollution generation in region or nation A to ecological damage in region or nation B is long, misdirected actions and even inaction have additional serious consequences. The time for correcting serious ecological damage is correspondingly long. The clear deduction is: Sound public policy must be based on a thorough understanding of excruciatingly complex problems, and increased research is a critical part of the information base need for decisions in environmental policy.

For the United States, the crux of the problem is that economic growth and environmental quality are now both at risk. Clearly there are costs to environmental preservation. However, sound economics recognizes that calculating profit and loss must account for resource depletion and waste production and management. The regulatory environment alone is increasingly forcing chemistry-based industries to adapt new technology or close down environmentally unacceptable operations. Thus, economic competitiveness and environmental quality are intertwined -- they can be mutually addressed in a "win/win" approach or be mutually ignored -- inviting double losses.

The NSF-CCR Program in Environmentally Benign Chemical Synthesis and Processing

The National Science Foundation has taken a small step towards addressing a selected subset of environmental problems. The Division of Chemistry and the Division of Chemical and Transport Systems have joined with the Council for Chemical Research in sponsorship of a new program -- ***Environmentally Benign Chemical Synthesis and Processing*** (hereinafter referred to as "Benign Chemistry" or "Benign Manufacturing")(3).

The program is designed to support pre-competitive research projects in chemical engineering and chemistry aimed at reducing pollution at its source. Active participation by researchers both in universities and in the many industries whose manufacturing processes depend on chemistry is required. Thus, the impact includes chemicals, petro- and agrochemicals, pharmaceuticals, synthetic materials, electronics, and energy -- quite a range of science and technology.

Industry is already marching ahead in its own interest to reduce waste and emissions through good economics and good housekeeping -- controlling fugitive emissions by proper attention to tanks, valves, and the like. Fundamental research is not needed to speed improvements of this type. The focus of the "Benign Chemistry" program is fundamental research in chemistry and engineering that can make a difference by leading to the discovery and development of advanced, environmentally benign methods for chemical synthesis and processing in industry.

This joint initiative of the NSF and the CCR is intended to finance long-range research supporting new synthetic methodologies, novel process designs and process improvements that reduce the potential for environmental release in the first place. Projects are sought which will lead to feedstock substitutions, alternative synthetic and separation procedures, more specific and efficient catalysts, and process modifications that would minimize byproduct formation and reduce waste production at the source. Industrial participation in the design, evaluation, and conduct of such research will ensure the viability of proposed projects in real-world manufacturing situations.

Some of the generic research areas in which eligible projects might fall include, but are not limited to, the following:

- new chemistries and methodologies for on-demand, on-site production and consumption of toxic chemical synthesis intermediates;
- new, more highly selective catalysts to increase product yield and reduce byproduct formation;
- low-energy separation technologies for feedstock purification and recycling;
- alternative chemical syntheses that bypass toxic feedstocks and solvents;
- improved membrane or molecular sieve technologies that integrate transport and reaction to enhance specificity;
- new processing methods that eliminate hard-to-entrap submicron-sized aerosols;
- alternative chemical syntheses that eliminate or combine process steps;
- novel low-temperature or other energy-efficient methods for synthesis and processing.

There are already many examples of new technologies which have combined good economics with "green" science and engineering. Here are just a few of them -- taken from a report of the World Resources Institute (4).

- Replacement of organic solvents by water in several manufacturing steps in the pharmaceutical industry has reduced volatile organic emissions by 100% and has repaid the engineering cost for the new process by savings within one year.
- In the microelectronics industry, replacement of caustics by vibratory cleaning reduced sludge 100% and was paid back within 2 years.
- In the photographic industry, incorporation of ion exchange for electrolyte recovery saved 85% of developer and 95% of silver; again -- payback within one year.

Projects must show appropriateness to current national concerns for pollution reduction or prevention; vague arguments that the proposed research may eventually reduce pollution are not compelling. Nonetheless, research must be fundamental in nature. In other words, the program is really seeking innovative and high risk/high payoff ideas. It does not invite studies of "the problem," but rather specific approaches to possible solutions. The initiative is not about such things as remediation or catalysis for reaction rate enhancement alone. In short, this research program seeks solutions to pollution prevention and reduction, not improvements in waste treatment.

The program is explicitly designed to encourage interdisciplinary interactions of all kinds. The idea is to stimulate the breakthroughs that come with cross fertilization by research partners of different backgrounds and perspectives. It aims to stimulate partnerships not only between industry and academe, but also between chemists and engineers, or between chemists, engineers and microbiologists, ecologists, or whoever has the needed complementary expertise.

Industrial participation, however, is a requirement. The participation most needed here -- to ensure relevance to an industrial setting and which is the *sine qua non* -- is intellectual partnership, which must be detailed in the proposal. Financial participation and other services in kind by the industrial partner are clearly welcomed, but not required. In fact, they will follow naturally in most cases where the intellectual partnership is real.

The Council for Chemical Research plays a critical role in this regard. It has agreed to act as a broker in this research program to assist university researchers in developing the necessary liaisons with appropriate industrial counterparts. Because industry representatives to the CCR are typically at the Research Director level, university-based engineers and chemists therefore enjoy access to industrial partners and problems at the highest levels.

Member companies of the Council for Chemical Research have also taken the initiative to generate a list of ideas -- possible generic projects which are ripe for collaboration and which could be of interest to academic scientists who would like to turn their talents to environmental problems of national importance, but who may not know which specific problems or industries to target.

The ideas already put forth by potential industrial partners are impressive in their range. The list is also impressive in illustrating how many gaps there still are in our understanding of many chemistry fundamentals.

- For example, to replace chromium in corrosion protection, one must develop some new redox chemistry.
- To recycle rubber more effectively, one could look for new ways to reverse crosslinking and vulcanization.
- Major reductions in pollution could be achieved by replacing traditional acid and base catalysts in bulk processes, perhaps using new zeolites.
- New water-based synthesis and processing methods are needed to minimize use of volatile organic solvents.
- Entirely new catalytic processes -- based on light, sound or antibodies -- might replace traditional heavy-metal ones. Alternatively, one might devise better chelates to separate and recycle heavy metals.

Clearly, such chemistry problems require more than mere tinkering with current knowledge and technology: they raise profound questions about which a lot of basic research is still needed. (See Chapter 1 for additional problems.)

1992 was the first year of the Environmentally Benign Synthesis and Processing Program. The announcement was distributed in February and the first proposals came in during the Spring to Chemical Engineering, stimulated by an April 10 deadline. Twenty-five proposals were received and eight awards were made, providing about $1 million in research support. Two of the projects involved collaborations between chemists and engineers as well as between industry and academe, and were funded jointly by the two cooperating NSF Divisions. Among the funded projects were ones on oil-water interfacial synthesis as an alternative to traditional solvent-based routes; new methods for sulfolane synthesis, including computer-assisted synthetic design; zeolite-based catalysis in alkylation of isoparaffins with olefins; new methods for removing flux residues from printed wiring; production of crosslinked polymer films through cationic photopolymerization; electrolytic steelmaking; and others. Clearly, as hoped, clever individuals generated ideas unforeseen by either NSF or the industrial idea-generators.

In 1993 and 1994, proposals were submitted to Chemistry as well as to Chemical Engineering. Chemistry Division programs, primarily in organic and in inorganic synthesis, have funded eleven projects totaling nearly $3 million. Catalytic, bio-catalytic, and photo-catalytic routes to organics are being tested by many of these investigators, while others are attempting reaction chemistry in environmentally "friendlier" solvents, such as water or supercritical CO_2. One investigation concerns computer modeling of alternative reaction paths and another seeks improved catalyst immobilization and regeneration.

A Broader View of Environmental Chemistry

What about the future? The research challenges to chemists in environmental chemistry are not limited to benign chemical synthesis, though synthesis and processing are clearly the base of the internationally profitable chemical industry. Chemistry is also central to improved detection, identification, monitoring, and separation of natural pollutants as well as industrial byproducts; to understanding important catalytic processes which occur on aerosols; to elucidating the mechanisms of photochemical changes from smog production to pesticide biodegradation. Chemists will develop the new catalysts for energy- and resource- efficient synthesis and will devise the biomimetic schemes for energy harvesting and bioremediation. Chemists will invent new disposal strategies for hazardous wastes of all types -- through encapsulation, biodegradation, and incineration. There are important, yet unsolved problems of basic research underlying all of these challenges.

In 1988 John Frost (then at Stanford University) and David Golden (SRI) chaired an NSF workshop which produced a report on Chemistry and the Environment (5). That report represented the "first salvo" of the Chemistry Division's efforts both to engage the chemistry community in environmental science and to alert relevant audiences in the government that resources would be needed. The Frost-Golden report set forth some of the important ways in which chemists could contribute their skills to mitigating problems of pollution in the soil, water, and air. The report pointed out as well that there was an essential basic research component needed for environmental science -- a role which could not be completely filled by the EPA or other agencies and which must be filled, and quite appropriately, by NSF.

In March 1992, the NSF Chemistry Division sponsored another workshop on environmental chemistry. This study was chaired by Thom Dunning (Battelle Northwest Laboratory) and Tom Spiro (Princeton), and took a somewhat different tack. It aimed to identify chemistry with environmental science by highlighting those familiar areas in which opportunities to advance chemistry's intellectual frontiers intersect with solutions of environmental problems. In other words, participants pointed out where fundamental questions in catalysis or photochemistry or synthesis or electrochemistry impact questions of air pollution or ground water purity, etc. The discussion covered most areas of chemistry. The group also took a look at related issues:

- How can environmental issues be used effectively to enhance education of chemists and to train the next generation of chemists more broadly? Conversely, how can chemistry be used to enhance education and understanding of environmental problems so that our citizenry can make wise, informed decisions on issues in which scientific understanding is important?
- What modes of support will be required to sustain an effective program of research in environmental chemistry? What balance is required in the portfolio of grants to individuals, groups, or centers? What special activities are needed -- such as encouragement for instrument development incorporating new analytical techniques, or special workshops to stimulate industry-academe-government interactions? Should we implement a program of seed grants or grant supplements to let enterprising chemists explore environmental side streets?
- How can we promote the international cooperation which will be necessary to address certain global problems and which can help avoid costly duplicative efforts in many areas? What do we need to do to stimulate additional cross-disciplinary collaboration between chemists and engineers and between chemists and biologists, geologists, and other scientists?

The insights gained during workshop discussions have guided the scientific scope and structure of the Chemistry Division's recent activities in environmental chemistry, which reflected a $7.6 million investment in FY1994. The emphasis on important fundamental science continues, with a focus on individual and interdisciplinary group research. It is probably worth stating that this is emphatically not just "business as usual." In the Environmentally Benign Chemistry Program, we insist that proposal authors include an "Environmental Impact Statement," which describes how the proposed research is expected to contribute to the programmatic goal of reducing pollution. Perhaps it will be necessary to extend this stratagem explicitly to other environmental research projects -- insisting that the relationship of the fundamental chemistry questions posed and the environmental problem which they would alleviate be spelled out.

Environmental Science at NSF

Environmental research at NSF spans many scientific disciplines, and chemistry's program is part of this broader effort. Overall, the goals of NSF's effort are (a) to build the science and engineering knowledge base for sound environmental decisions; (b) to develop new instruments, technologies, and databases to undergird environmental research; (c) to develop human resources to tackle complex, multi-disciplinary problems; and (d) to develop partnerships with academe, industry, and environmental policy-makers.

The scope of this effort is approximately $300 million per year, about half of which is devoted to Global Change research. Research in biodiversity and ecosystem dynamics accounts for nearly a quarter of total expenditures. Although Chemistry is a relatively small participant, it is part of a rapidly growing interest in environmental technology within NSF, and is linked to environmentally conscious manufacturing, mitigation of the effects of toxic substances and natural disasters, and improvement of air and water quality.

"Environmental technologies" are broadly understood to include hardware, software, processes, systems, and services which result in environmentally conscious production, use, or disposal of products and substances. The overarching goal of fundamental research on environmental technologies is to enable sustainable development which will allow U.S. manufacturing, including chemical manufacturing, to grow while reducing environmental insult and adverse health effects.

Developing alternative manufacturing practices for the chemical industry will demand not only clever new synthesis and processing methods, but a better understanding of the behavior and effects of chemicals in the environment. Interdisciplinary research not only with chemical engineers, but also with geoscientists and ecosystem dynamics experts, will be necessary to obtain a full understanding of environmental impact and to construct effective environmental regulations. The demand for chemists to develop multi-disciplinary and multi-sector partnerships can only be expected to increase.

Partnership between NSF and EPA

Another development in environmental chemistry research occurred in 1993 with the signing of a memorandum of understanding (MOU) on pollution prevention between NSF's Chemistry Division and EPA's Office of Pollution Prevention and Toxics. The purpose of the MOU is to coordinate EPA's Design for the Environment programs with NSF's Environmentally Benign Synthesis programs and to foster a new environmental consciousness among chemists, especially those involved in synthesis. Joint efforts have focused on informing chemists about the opportunities for fundamental research in environmental chemistry, primarily through symposia. Precedent has also been set for joint funding and review of research projects by the two agencies. Cooperation of this type between a research agency and a regulatory agency signals a more integrated Federal approach to solving environmental problems.

Conclusion

A summary might come from two different points of view.

From the viewpoint of thinking about chemistry-based manufacturing and the environment, it becomes clear that what we need are paradigm shifts. We need to move from an atmosphere of monitoring to minimize pollution to one in which environmentally benign manufacturing is the norm -- because it makes economic, engineering, and environmental sense by conserving both natural and man-made resources. We must shift from an atmosphere of regulation and litigation to one of designed minimization of environmental impact. We must move from a mindset of clean-up after the fact to designed, built-in recyclability and biodegradability.

In short, for chemists the environment and its preservation present formidable scientific and technical challenges. Linked to these are wonderful scientific and engineering research problems of stimulating complexity. Addressing the scientific problems of the environment through environmental chemistry will be intellectually satisfying. But it will also be satisfying in another way -- by being part of the solution instead of part of the problem. And for today's brightest students who want to be challenged intellectually, to look at problems from a broad, interdisciplinary perspective, and to contribute to society -- what could be better than the interdisciplinary training provided through research and education in environmental chemistry?

Any opinions, findings and conclusions or recommendations expressed in this publication are those of the authors and do not necessarily reflect the views of the National Science Foundation.

Literature Cited

1. *Chemical and Engineering News* **1994 (July 4)**, 72 (27), pp. 54-60.
2. Suter II, G. W. *Ecological Risk Assessment*; Lewis Publishers: Ann Arbor, MI, 1993.
3. *Environmentally Benign Chemical Synthesis and Processing: Research on Pollution Prevention at its Source;* [NSF 92-13]; National Science Foundation: Washington, DC, 1992.
4. Huisingh, D. *Good Environmental Practices -- Good Business Practices;* Wissenschaftentrum Berlin fur Sozialforschung: Berlin, Germany, 1988, cited in Heaton, G.; Repetto, R.; Sobin, R. *Transforming Technology: An Agenda for Environmentally Sustainable Growth in the 21st Century;* World Resources Institute: New York, NY, 1992.
5. *Chemistry and the Environment - 1988: A National Science Foundation Workshop Report*; Frost, J. W.; Golden, D. M., Eds.; Stanford, CA, 1988.

RECEIVED September 16, 1994

BENIGN CHEMISTRY: RESEARCH

Chapter 3

Microbial Biocatalysis
Synthesis of Adipic Acid from D-Glucose

Karen M. Draths and John W. Frost

Department of Chemistry, Michigan State University, East Lansing, MI 48824

Four billion pounds of adipic acid are produced each year using petroleum-based feedstocks, carcinogenic benzene as starting material, and extreme reaction conditions. Nitrous oxide, which plays a role in ozone layer depletion, is emitted as a byproduct. As an alternative to the currently employed synthetic methodology, a two-step synthesis of adipic acid from D-glucose has been developed which eliminates each of these problems. A microbial catalyst was created which possesses a novel biosynthetic pathway that synthesizes *cis, cis*-muconic acid from D-glucose. This pathway does not occur in nature but has been created in a strain of *Escherichia coli*. *Cis, cis*-muconic acid is exported to the culture supernatant, where it is hydrogenated under mild conditions to yield adipic acid.

Four billion pounds of adipic acid are synthesized each year (*1*) using synthetic methodology that is likely to be in conflict with increasingly stringent environmental regulations. A small percentage of the adipic acid which is produced is used in preparation of polyurethanes, lubricants, and plasticizers and as a food acidulant (*1*). However, nearly all of the adipic acid that is synthesized is used to satisfy an 8.8 billion pound annual demand for nylon-6,6 (*2*), a polyamide produced by condensation and polymerization of adipic acid with hexamethylenediamine (*2*). While uses (*2*) for nylon-6,6 have expanded from women's nylon hosiery to include carpet fibers, upholstery, apparel, tire reinforcements, and various types of plastic, the basic synthetic methodology for adipic acid manufacture remains largely unchanged since its development in the early 1940's (*1*).

Synthetic methodology accounting for nearly all of the adipic acid synthesized in the United States (*3*) is outlined in Figure 1. Benzene is hydrogenated to produce cyclohexane (*4,5*), which is subsequently air oxidized in the presence of metal catalysts to yield a mixture of cyclohexanone and cyclohexanol (*1*). Nitric acid oxidation of the mixture affords adipic acid in high yields (92-96%) with decarboxylation resulting in glutaric and succinic acid formation constituting the major byproducts (*1*). Du Pont, a major producer of adipic acid, markets the dimethyl esters of adipate, glutarate, and succinate as an alternative to methylene chloride for paint stripping formulations. A second process that accounts for a small percentage of adipic acid synthesis employs hydrogenation of phenol to obtain predominantly cyclohexanol (*3*). Oxidation with nitric acid yields adipic acid (*1*).

Nitrous oxide, a byproduct of nitric acid oxidation of cyclohexanone and cyclohexanol, is released to the atmosphere during adipic acid production (1,3). Due to the massive scale on which it is carried out, adipic acid manufacture accounts for some 10% of the annual increase in atmospheric nitrous oxide levels (6). Nitrous oxide is a causative agent of atmospheric ozone depletion (6) and has also been implicated in the global warming phenomenon known as the greenhouse effect (7). Large-scale release of nitrous oxide into the atmosphere will not likely be tolerated in the near future (3). Although Du Pont is developing technology to remove nitrous oxide from gaseous effluents chemically, the cost of completely removing nitrous oxide has not been established (8).

Chemical synthesis at the cost of generating a hazardous byproduct is not unique to adipic acid production. This and other problems characteristic of the chemical industry are illustrated by further examination of adipic acid synthesis. Benzene, the primary starting material in adipic acid manufacture, is a proven carcinogen (9). Benzene is used widely in the chemical industry, particularly as a feedstock (4). For example, benzene is used to make phenol (4), the starting material from which a small percentage of adipic acid is currently synthesized. The United States alone produced over 12 billion pounds of benzene in 1993 (10). Benzene is derived exclusively from petroleum (4,5), a non-renewable fossil fuel. Of the chemicals in the United States which are produced in excess of 10 million pounds per year, 98% are derived from petroleum feedstocks (4). Finally, extreme reaction conditions which are used in adipic acid manufacture include temperatures up to 250°C and pressures which reach 800 psi. Reaction conditions such as these are used routinely by the chemical industry.

Increased sensitivity to the effect of manufacturing on the environment will require fundamental change on the part of the chemical industry, which until recently has focused solely on producing a chemical for the lowest price possible. Under a different accounting system that is currently being discussed, full-cost accounting (11), chemical prices would reflect the cost of using a nonrenewable, petroleum-derived feedstock instead of a renewable feedstock. The environmental impact of a hazardous byproduct and costs associated with exposure of industry employees to carcinogens would also be reflected (11). In addition, the new Chemical and Manufacturing Rule issued by the U.S. Environmental Protection Agency requires reduction of more than 100 hazardous organic air pollutants, including benzene, by up to 88% (12,13,14). The EPA estimates that the Chemical and Manufacturing Rule, which will be phased in over the next three years, will cost $450 million in capital expenditures and will increase annual operating expenses of the chemical industry by $230 million. Environmental regulations such as these and consumer pressure will likely provide powerful incentives for the chemical industry to develop environmentally benign syntheses of chemicals.

In addition to the currently used technology, small scale syntheses of adipic acid which use traditional chemical methodology (1,15) or biological processes (16,17,18,19,20) have been developed over the years. However, none of these processes simultaneously address the problems of petroleum-based feedstocks, carcinogenic starting materials, generation of environmentally hazardous byproducts, and the use of forcing reaction conditions. An environmentally compatible synthesis of adipic acid from D-glucose (Figure 1) was recently elaborated (21) which eliminates each of these problems encountered in current adipic acid production. D-Glucose is a nontoxic sugar derived from plant starch and cellulose (22). Availability of D-glucose compares favorably to petroleum since additional feedstock can be generated in the fields during each growing season. Ozone-depleting gases and greenhouse gases are not generated. By virtue of reliance on enzyme-catalyzed conversions, mild reaction temperatures and pressures are utilized.

The two-step synthesis of adipic acid utilizes a biocatalyst to convert D-glucose into *cis, cis*-muconic acid, which is subsequently hydrogenated under mild

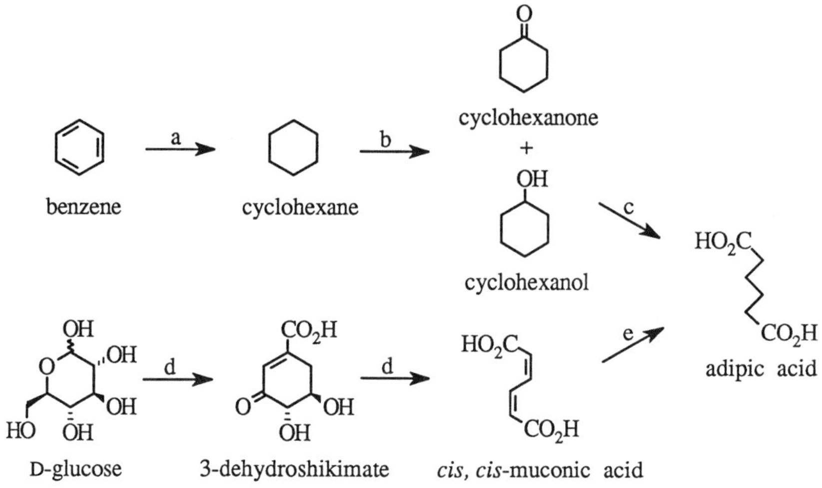

Figure 1. Comparison of current industrial synthesis of adipic acid from benzene (a-c) to synthesis of adipic acid from D-glucose (d, e). (a) Ni-Al$_2$O$_3$, 370-800 psi, 150-250°C. (b) Co, O$_2$, 120-140 psi, 150-160°C. (c) Cu, NH$_4$VO$_3$, 60% HNO$_3$, 60-80°C. (d) *E. coli* AB2834*aroE*/pKD136/pKD8.243A/pKD8.292. (e) 10% Pt on carbon, H$_2$, 50 psi. (Adapted and reproduced with permission from ref. 21.)

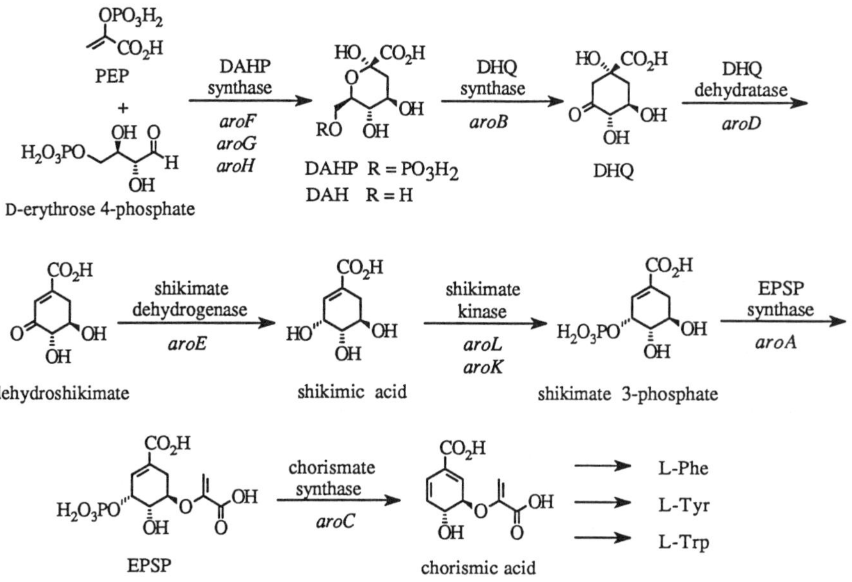

Figure 2. The common pathway of aromatic amino acid biosynthesis. Enzymes and the *E. coli* loci which encode the enzymes are indicated.

reaction conditions. The microbial catalyst is a strain of *Escherichia coli* that possesses a biosynthetic pathway which does not occur in nature. Enzymes associated with the pentose phosphate pathway, aromatic amino acid biosynthesis, hydroaromatic catabolism, and the benzoate and *p*-hydroxybenzoate branches of the β-ketoadipate pathway constitute this novel biosynthetic pathway. Development of the microbial catalyst proceeded through several stages. Since an economical synthesis requires maximum conversion of starting material into product, the biocatalyst was designed to direct as much D-glucose as possible into aromatic amino acid biosynthesis, which constitutes the front portion of *cis, cis*-muconate biosynthesis. It was then necessary to ensure that all the D-glucose in the pathway was converted to 3-dehydroshikimate (Figure 1), the intermediate from which *cis, cis*-muconate biosynthesis diverges from aromatic amino acid biosynthesis. Finally, localization of genes from other bacterial organisms in *E. coli* was needed for expression of three enzymes which are not normally found in the host strain. Those enzymes, 3-dehydroshikimate (DHS) dehydratase, protocatechuate decarboxylase, and catechol 1,2-dioxygenase, were necessary for conversion of 3-dehydroshikimate into *cis, cis*-muconic acid.

Choice of *E. coli* as Microbial Catalyst

It was necessary at the outset to choose a bacterial organism to act as host for the newly created biosynthetic pathway. *Escherichia coli* was selected because of the extensive biochemical and molecular biological databases which exist for this organism (*23*). Advantage was taken of libraries of *E. coli* genes and auxotrophic mutants for development of a catalyst which efficiently synthesized 3-dehydroshikimate from D-glucose. Expression of DHS dehydratase, protocatechuate decarboxylase, and catechol 1,2-dioxygenase activities from foreign genes localized in *E. coli* would also be necessary. Fortuitously, *E. coli* is the paradigm in which protein expression has been exhaustively examined. Another reason for choosing *E. coli* as the host follows from the likely need for optimization of the microbial catalyst for commercial synthesis of adipic acid. Extensive biochemical and molecular biological data accrued on *E. coli* will enhance this developmental process.

Direction of D-Glucose into Aromatic Biosynthesis

A critical aspect to the design of the microbial catalyst for *cis, cis*-muconate synthesis is to direct a high percentage of the carbon (in the form of D-glucose) consumed by the microbe into the common pathway of aromatic amino acid biosynthesis. This pathway (Figure 2) (*24*) is found in plants, bacteria, and fungi, and is responsible for conversion of phosphoenolpyruvate and D-erythrose 4-phosphate into chorismic acid, the central intermediate from which the aromatic amino acids and related secondary metabolites are synthesized. Increasing the flow of carbon into the common pathway has traditionally been achieved by increasing the catalytic activity of the first enzyme, 3-deoxy-D-*arabino*-heptulosonic acid 7-phosphate (DAHP) synthase (*25*). This has been accomplished in several ways. Localization of a gene which encodes one of the three isozymes of DAHP synthase on an extrachromosomal plasmid (*26*) or removal of transcriptional control of one of the DAHP synthase genes (*27,28*) results in elevated enzyme activity due to an increased quantity of protein in the cell. Alternatively, introduction of a mutation in one of the DAHP synthase genes can render the protein resistant to feedback inhibition by aromatic amino acids (*29,30*).

We quantified DAHP production by various *E. coli aroB* strains in order to understand the effect of certain parameters on direction of carbon flow into the common pathway (*31,32*). A mutation in the gene (*aroB*) encoding 3-dehydroquinate (DHQ) synthase leaves *E. coli aroB* strains incapable of converting

DAHP into DHQ. In bacterial metabolism, the substrate of a missing enzyme is frequently exported to the culture supernatant. In this case, the vast majority of the substrate is dephosphorylated and exported as 3-deoxy-D-*arabino*-heptulosonic acid (DAH) (*26*). DAH is readily quantitated (*33*) in the culture supernatant and provides a measure of the carbon flow directed into the common pathway. The effect of increased DAHP synthase activity on carbon flow into the common pathway was demonstrated by examination of *E. coli* AB2847*aroB*/pRW5 (Figure 3) (*32*). Plasmid pRW5 contains a feedback-resistant copy of the *aroG*-encoded DAHP synthase under transcriptional control of the *lac* promoter. Increasing the concentration of transcriptional inducer isopropyl β-D-thiogalactopyranoside (IPTG) from 0 to 50 mg/L results in a sixfold increase in DAHP synthase specific activity (Figure 3B). Similarly, the concentration of DAHP which is synthesized increases approximately fivefold (Figure 3A). Further increases in DAHP synthase activity, however, appear to have no net impact on delivery of carbon to aromatic biosynthesis, as demonstrated by examination of *E. coli* AB2847*aroB*/pRW5/pRW300 (*32*). At the maximum level of induction, an additional high copy plasmid (pRW300) containing the *aroG* gene results in a further sixfold elevation in DAHP synthase specific activity relative to AB2847*aroB*/pRW5 (Figure 3B). However, no increase in DAHP synthesis is detected (Figure 3A).

Recognition that a substrate of DAHP synthase may have a role in limiting direction of metabolites to the common pathway was critical to improving the overall conversion of D-glucose into product. D-Erythrose 4-phosphate was an excellent candidate for this limiting role. Although its intermediacy in cellular metabolism is well-established, D-erythrose 4-phosphate has never been convincingly detected in a living system (*34,35*). Consideration of which enzyme(s) might have a role in determining cellular availability of D-erythrose 4-phosphate led to the enzyme transketolase. As part of the non-oxidative branch of the pentose phosphate pathway (*36*), transketolase and transaldolase siphon D-glucose equivalents from glycolysis into several biosynthetic pathways. The catalytic interplay of transketolase and transaldolase leads to generation of D-ribose 5-phosphate, D-sedoheptulose 7-phosphate, and D-erythrose 4-phosphate, which, respectively, are required for biosynthesis of nucleotides, lipopolysaccharides, and aromatic amino acids.

Although both transketolase and transaldolase play a role in D-erythrose 4-phosphate generation, attention was focused on increasing transketolase activity for the following reasons. Transketolase catalyzes the reversible transfer of a two carbon ketol group between various aldose acceptors (Figure 4). The equilibrium constant for transketolase-catalyzed conversion of D-fructose 6-phosphate into D-erythrose 4-phosphate favors D-fructose 6-phosphate (*37*). D-Erythrose 4-phosphate may actually be "masked" as D-fructose 6-phosphate to circumvent the intractable chemical characteristics of free D-erythrose 4-phosphate (*38*). Notably, *E. coli* mutants which possess significantly reduced levels of transketolase are incapable of growth in the absence of aromatic amino acid supplementation, presumably due to limited D-erythrose 4-phosphate availability (*39*). Recent measurements of flux control coefficients in human erythrocytes have likewise indicated that transketolase exerts significantly more regulatory control on the pentose phosphate pathway than does transaldolase (*40*).

An *E. coli* gene encoding transketolase (*tkt*) was isolated from a genomic library by complementation of a mutant possessing reduced transketolase levels (*32*). Subcloning of the original insert resulted in localization of the *tkt* gene to a 5 kb *Bam*H I fragment. Purification of transketolase to homogeneity from both wild-type *E. coli* and from an overexpressing strain containing the 5 kb fragment on an extrachromosomal plasmid revealed that the protein is a homodimer with a subunit molecular weight of 72 kDa (*32*). This is in close agreement with nucleotide sequence information (*41*). Recently, a second transketolase gene (designated *tktB*) was isolated from *E. coli*, and the nucleotide sequence was determined (*42*). This

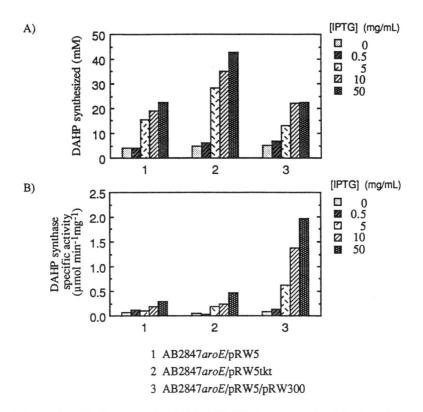

Figure 3. (A) Concentration (mM) of DAH that accumulated in the culture supernatants of *E. coli aroB* strains after 21 h of growth. (B) Specific activity of DAHP synthase (μmol min^{-1} mg^{-1}) after 21 h of growth. DAH concentrations and DAHP synthase specific activities were determined for each construct at IPTG concentrations of 0, 0.5, 5, 10, and 50 mg/L. Constructs examined included AB2847*aroB*/pRW5 (strain 1), AB2847*aroB*/pRW5tkt (strain 2), and AB2847*aroB*/pRW5/pRW300 (strain 3). (Adapted and reproduced with permission from ref. 32.)

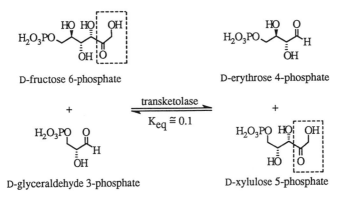

Figure 4. Transketolase-catalyzed removal of a two carbon ketol group (boxed) from D-fructose 6-phospate results in formation of D-erythrose 4-phosphate.

second transketolase isozyme accounts for only a small percent of wild-type transketolase activity (*42*).

The effect of transketolase activity on delivery of D-glucose to aromatic biosynthesis was examined by comparison of DAHP synthesis by AB2847*aroB*/pRW5 and AB2847*aroB*/pRW5tkt (*32*). These strains possess comparable levels of DAHP synthase activity at identical concentrations of IPTG (Figure 3B). What distinguishes AB2847*aroB*/pRW5tkt from AB2847*aroB*/pRW5 is an extrachromosomal copy of *tkt* localized on plasmid pRW5tkt which results in elevated transketolase specific activity. Under the same culturing conditions, AB2847*aroB*/pRW5tkt synthesizes twice the DAHP concentration relative to AB2847*aroB*/pRW5 (Figure 3A). This demonstrates that increases in transketolase activity resulted in higher D-erythrose 4-phosphate availability and increased flow of carbon into the common pathway of aromatic amino acid biosynthesis.

Alleviation of Rate-Limiting Enzymes

The next step in designing an efficient catalyst was to take full advantage of the increased carbon flow directed into aromatic biosynthesis caused by elevated transketolase and DAHP synthase activities. This demands that all of the carbon committed to aromatic biosynthesis be converted into 3-dehydroshikimate (DHS), the common pathway intermediate from which *cis, cis*-muconate biosynthesis would diverge. A mutation in the gene (*aroE*) encoding shikimate dehydrogenase results in accumulation of DHS, the substrate of the missing enzyme, in the culture supernatant. Analysis of the culture supernatant of *E. coli* AB2834 *aroE* by proton nuclear magnetic resonance (^{1}H NMR) indicated that this strain synthesized 9 mM DHS from 56 mM D-glucose (*43*). Since no other common pathway metabolites were detected in the culture supernatant, it was concluded that all the D-glucose directed into aromatic biosynthesis was converted into DHS.

To increase the direction of D-glucose into aromatic biosynthesis, plasmid pKD130A was prepared which contained genes encoding transketolase (*tkt*) and DAHP synthase (*aroF*) (*31*). ^{1}H NMR analysis of the culture supernatant of AB2834*aroE*/pKD130A grown under identical conditions as AB2834*aroE* indicated that 25 mM DHS and 9 mM DAH were synthesized (*42*). Given the increased concentration of metabolites in the pathway, wild-type levels of DHQ synthase are not sufficient to catalyze conversion of DAHP into DHQ. The cell responds to increasing intracellular DAHP concentrations by exporting some as DAH to the culture supernatant. Once DAH has been exported out of the cytosol, it is lost to cellular metabolism, and the percent conversion of D-glucose into DHS is reduced. Likewise, the appearance of DAH in the supernatant compromises the purity of the product of the microbial catalyst.

To improve percent conversion and product purity, it was essential to remove the rate-limiting character of DHQ synthase. Plasmid pKD136 (*42*) was created by insertion of a 1.65 kb *aroB* fragment into the transketolase and DAHP synthase-encoding plasmid pKD130A. ^{1}H NMR analysis of the culture supernatant of AB2834*aroE*/pKD136 indicated that the strain synthesized 30 mM DHS from 56 mM D-glucose (*42*). No other intermediates from the common pathway were detected in the culture supernatant. Because DHQ was never detected in the culture supernatant, wild-type levels of DHQ dehydratase were apparently adequate to prevent intracellular accumulation and subsequent exportation of DHQ.

D-Glucose to *cis, cis*-Muconate Biosynthesis

Catalyzed conversion of D-glucose into *cis, cis*-muconic acid (*21*) required creation of a biosynthetic pathway not known to exist naturally (Figure 5). This pathway relied on DHS dehydratase (Figure 5, enzyme A) (*44,45*) to couple aromatic biosynthesis to

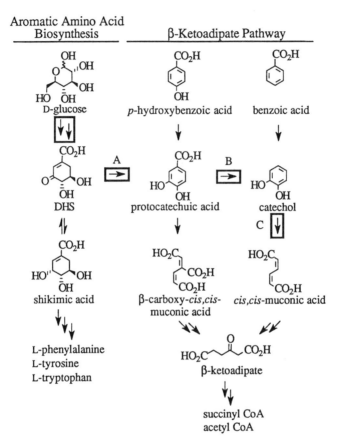

Figure 5. The biocatalytic pathway (boxed arrows) created for microbial conversion of D-glucose into *cis, cis*-muconate from the perspective of the biochemical pathways from which the enzymes were recruited. Conversion of D-glucose into DHS requires transketolase (*tkt*) from the pentose phosphate pathway and DAHP synthase (*aroF, aroG, aroH*), DHQ synthase (*aroB*), and DHQ dehydratase (*aroD*) from the common pathway of aromatic amino acid biosynthesis. Conversion of DHS into catechol requires DHS dehydratase (*aroZ*, enzyme A) from hydroaromatic catabolism, protocatechuate decarboxylase (*aroY*, enzyme B), and catechol 1,2-dioxygenase (*catA*, enzyme C) from the benzoate branch of the β-ketoadipate pathway. (Adapted and reproduced with permission from ref. 21.)

aromatic catabolism. In various fungi and bacteria, DHS dehydratase plays a critical role in allowing an organism to metabolize hydroaromatics such as shikimic acid as a sole carbon source. Conversion of DHS into protocatechuic acid catalyzed by DHS dehydratase yields a substrate which can be further metabolized via the *p*-hydroxybenzoate branch of the β-ketoadipate pathway (*46*). However, organisms which possess the ability to metabolize protocatechuic acid normally do so via β-carboxy-*cis, cis*-muconic acid (*46*). In order to properly channel protocatechuic acid into *cis, cis*-muconate biosynthesis, protocatechuate decarboxylase (Figure 5, enzyme B) was required. By catalyzing the conversion of protocatechuic acid into catechol, this enzyme (*47,48*) would essentially fuse the *p*-hydroxybenzoate branch to the benzoate branch of the β-ketoadipate pathway. Expression of catechol 1,2-dioxygenase (Figure 5, enzyme C) (*49*), which catalyzes conversion of catechol into *cis, cis*-muconic acid, completes the novel biosynthetic pathway.

Since DHS dehydratase, protocatechuate decarboxylase, and catechol 1,2-dioxygenase are not found under normal conditions in *E. coli*, it was necessary to express these enzymes from foreign genes in the microbial catalyst. The appropriate genes were isolated from organisms which catabolize hydroaromatics and aromatics as sole carbon sources via the β-ketoadipate pathway. Although the DHS dehydratase gene has been isolated from *Neurospora crassa* and *Aspergillus nidulans*, neither has been successfully expressed in *E. coli*. Attention was instead focused on *Klebsiella pneumoniae* as an alternative source of the DHS dehydratase gene. Previous experience with a gene encoding another enzyme involved in hydroaromatic catabolism indicated that a *K. pneumoniae* gene was readily expressed from its native promoter in *E. coli* (*50*). This may reflect the close evolutionary relationship (*51*) which exists between *K. pneumoniae* and *E. coli*. The gene encoding DHS dehydratase (*aroZ*) was localized to a 3.5 kb *Bam*H I fragment that was isolated from a genomic library of *K. pneumoniae* (*21*).

The role of protocatechuate decarboxylase in bacterial metabolism remains to be established (*52*). Although it was originally proposed as an essential component of *p*-hydroxybenzoate catabolism (*47,48*), recent examination has shown this not to be the case (*52*). Detection of protocatechuate decarboxylase activity in liquid suspension cultures of *K. pneumoniae* indicated the gene (*aroY*) could also be isolated from this organism. The *K. pneumoniae aroY* gene was localized to a 2.3 kb *Hin*d III fragment (*21*). As part of a comprehensive investigation of the β-ketoadipate pathway in *Acinetobacter calcoaceticus*, Ornston and coworkers have shown that the catechol 1,2-dioxygenase gene (*catA*) is expressed in *E. coli* (*49*). A plasmid containing the *catA* gene of *A. calcoaceticus* was generously provided by the Ornston group.

Final assembly of the microbial catalyst required amplification of three native *E. coli* genes including *tkt*, *aroF*, and *aroB* and expression of three foreign genes including *aroZ* and *aroY* from *K. pneumoniae* and *catA* from *A. calcoaceticus*. To ensure plasmid compatibility and plasmid maintenance, three vectors possessing compatible replication origins and each encoding resistance to a different antibiotic were identified. Vector pBR325 (*53*) contains the pMB1 origin of replication. Insertion of the *tkt*, *aroF*, and *aroB* genes into pBR325 to produce pKD136 (*43*) results in a plasmid which confers resistance to ampicillin. Plasmid pSU19 (*54*), which confers resistance to chloramphenicol, contains the p15A replication origin. Localization of the 3.5 kb *aroZ* fragment and the 2.3 kb *aroY* fragment, respectively, into the *Bam*H I and *Hin*d III sites of pSU19 resulted in pKD8.243A. Expression of the *catA* gene of *A. calcoaceticus* required its positioning behind a vector-encoded promoter. This was achieved by localization of the 1.5 kb *Sal* I/*Kpn* I fragment obtained from pIB1345 (*49*) behind the *lac* promoter of pCL1920 to afford pKD8.292. Plasmid pCL1920 (*55*) contains the pSC101 origin of replication and confers resistance to spectinomycin.

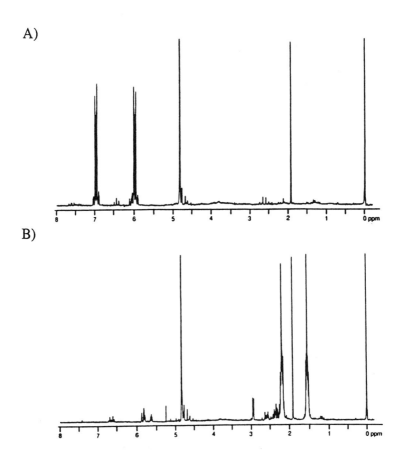

Figure 6. ^{1}H NMR (200 MHz) spectrum before (A) and after (B) hydrogenation of the unpurified, crude culture supernatant of *E. coli* AB2834*aroE*/pKD136/pKD8.243A/pKD8.292. Relevant resonances include those for (A) *cis, cis*-muconic acid δ 6.0 (2H), δ 7.0 (2H), and (B) adipic acid δ 1.5 (4H), δ 2.2 (4H). (Reproduced with permission from ref. 21.)

Plasmids pKD136, pKD8.243A, and pKD8.292 were transformed into *E. coli* AB2834*aroE*. By virtue of a mutation in the gene which encodes shikimate dehydrogenase (*aroE*), AB2834 is unable to catalyze conversion of DHS into shikimate. Elimination of shikimate dehydrogenase catalytic activity diverts all of the carbon flowing through the common pathway away from biosynthesis of aromatic amino acids and into the pathway designed for *cis,cis*-muconate synthesis. To determine expression levels of the foreign genes in the *E. coli* host, DHS dehydratase (*56*), protocatechuate decarboxylase, and catechol 1,2-dioxygenase (*57*) specific activities were measured in AB2834*aroE*/pKD136/pKD8.243A/pKD8.292. The specific activity of catechol 1,2-dioxygenase (0.25 units/mg) was significantly higher than the specific activities of DHS dehydratase (0.078 units/mg) and protocatechuate decarboxylase (0.028 units/mg).

Evaluation of AB2834*aroE*/pKD136/pKD8.243A/pKD8.292 by ^{1}H NMR analysis of the culture supernatant indicated that on a 1 liter scale in laboratory shake flasks, the catalyst converted 56 mM D-glucose into 16.8 mM ± 1.2 *cis, cis*-muconate (Figure 6A) (*21*). In order to accumulate *cis, cis*-muconate in the absence of *cis, trans*-muconate, the culture supernatant was maintained above pH 6.3 by periodic addition of sodium hydroxide. In the absence of sodium hydroxide addition, the culture pH dropped to approximately pH 5 within the first 24 h and *cis, cis*-muconate synthesis ceased due to the acidity of the solution. Prolonged incubation of *cis, cis*-muconate at acidic pH resulted in its complete isomerization to the *cis, trans* isomer. After addition of catalytic 10% platinum on carbon to the unpurified culture supernatant, the heterogeneous solution was hydrogenated at 50 psi for 3 h at room temperature. Reduction of *cis, cis*-muconate to adipic acid proceeded in 90% yield, as determined by ^{1}H NMR of the reaction (Figure 6B) (*21*).

Concluding Remarks

Before it will be practical to consider using a microbial catalyst to manufacture adipic acid, significant challenges must be addressed. To improve catalyst stability and increase the percent conversion of D-glucose into product, further development of the microbial catalyst will be needed. The scale of the reaction also requires adjustment from laboratory shake flasks to fermentation tanks, which would then be readily scaled to industrial production. Future efforts will focus on these challenges.

Even allowing for process optimization, synthesis of adipic acid from D-glucose using microbial catalysis does not compare favorably today from an economic standpoint with synthesis from benzene (Frost, J. and Lievense, J. *New J. Chem.*, in press). By making reasonable assumptions concerning process optimization and scale-up, production costs for adipic acid from D-glucose are estimated at $0.79/lb (Frost, J. and Lievense, J. *New J. Chem.*, in press), although advances in biocatalytic synthesis including co-factor regeneration would decrease the cost of production considerably (Frost, J. and Lievense, J. *New J. Chem.*, in press). Polymer-grade adipic acid currently sells for $0.65/lb (*58*). Considering that synthesis of adipic acid from benzene has been optimized and engineered over the past fifty years (*1*), the results of this comparison are not unexpected.

However, if one considers costs which might be factored in under full-cost accounting, the price of adipic acid production from benzene would likely change dramatically. For example, the cost of protecting Persian Gulf shipping lanes was recently estimated at $23.50 per barrel of petroleum (*59*), more than doubling the approximate $15 per barrel price of petroleum. The fact that petroleum is a nonrenewable resource should also be considered. Furthermore, the Environmental Protection Agency recently issued a document (*3*) designed to assist state agencies in implementation of nitrous oxide emission standards. Costs necessary to bring current production in line with such regulations would also be considered.

Biocatalytic synthesis is a field still in its infancy. Given that current commercial production of lactic acid ($0.71-1.15/lb) (*58*) and L-lysine ($1.20/lb) (*58*) employ biocatalytic routes, the view of what can be synthesized practically and economically from D-glucose is changing rapidly. As demonstrated by this report, there is no reason to limit biocatalytic methodology to production of metabolic end products such as amino acids and vitamins. Creation of a microbial catalyst constitutes only the first step toward providing an alternative method of adipic acid production. Further development is justified by the advantages derived from synthesis of adipic acid from D-glucose. These advantages include replacement of a nonrenewable feedstock with a renewable feedstock, elimination of volatile organics such as benzene and cyclohexane, avoidance of harsh reaction conditions, and elimination of nitrous oxide as a byproduct.

Acknowledgments

The authors thank Professor L. N. Ornston for plasmid pIB1345. Research was supported under the Design for the Environment Program administered by the Office of Pollution Prevention and Toxics at the Environmental Protection Agency, the National Science Foundation, and Genencor International, Inc.

Literature Cited

1. Davis, D.D.; Kemp, D.R. In *Kirk-Othmer Encyclopedia of Chemical Technology*; Kroschwitz, J.I.; Howe-Grant, M., Eds.; Wiley: New York, NY, ed. 4, 1991, Vol. 1; pp 466-493.
2. Putscher, R.E. In *Kirk-Othmer Encyclopedia of Chemical Technology*; Grayson, M.; Eckroth, D., Eds.; Wiley: New York, NY, ed. 3, 1982, Vol. 18; pp 328-371.
3. *Alternative Control Techniques Document - Nitric And Adipic Acid Manufacturing Plants*; U.S. Environmental Protection Agency. Office of Air and Radiation. Office of Air Quality Planning and Standards. U.S. Government Printing Office: Washington, DC, 1991; EPA-450/3-91-026.
4. Szmant, H.H. *Organic Building Blocks of the Chemical Insustry*; Wiley: New York, NY, 1989.
5. Franck, H.-G.; Stadelhofer, J.W. *Industrial Aromatic Chemistry*; Springer-Verlag: New York, NY, 1988.
6. Thiemens, M.H.; Trogler, W.C. *Science* **1991**, *251*, 932.
7. Dickinson, R.E.; Cicerone, R.J. *Nature* **1986**, *319*, 109.
8. *Chem. Eng. News* **1991**, *69*, 15 (March 11).
9. Lenga, R.E.; Votoupal, K.L. *The Sigma-Aldrich Library of Regulatory and Safety Data*; Sigma-Aldrich: Milwaukee, WI, 1993.
10. Reisch, M.S. *Chem. Eng. News* **1994**, *72*, 12 (April 11).
11. Popoff, F.P.; Buzzelli, D.T. *Chem. Eng. News* **1993**, *71*, 8 (January 11).
12. Noah, T. *Wall Street J.* **1994**, B2 (March 2).
13. *Washington Post* **1994**, A3 (March 2).
14. Ember, L. *Chem. Eng. News* **1994**, *72*, 4 (March 7).
15. Brookhart, M.; Sabo-Etienne, S. *J. Am. Chem. Soc.* **1991**, *113*, 2777.
16. Kaneyuki, H.; Ogata, K. U.S. Pat. 3 912 586, 1975.
17. Maxwell, P.C. U.S. Pat. 4 355 107, 1982.
18. Dainippon Ink and Chemicals, Inc. Jp. Pat. 82 129 694, 1982.
19. Faber, M. U.S. Pat. 4 400 468, 1983.
20. Nissan Chemical Industries, Ltd. Jp. Pat. 58 149 687, 1983.
21. Draths, K.M.; Frost, J.W. *J. Am. Chem. Soc.* **1994**, *116*, 399.
22. Lynd, L.R.; Cushman, J.H.; Nichols, R.J.; Wyman, C.E. *Science* **1991**, *251*, 1318.

23. *Escherichia coli and Salmonella typhimurium*, Neidhardt, Ed.; American Society for Microbiology: Washington, DC, 1987.
24. *The Shikimate Pathway*; Conn, E. E., Ed.; Plenum: New York, NY, 1974.
25. Herrmann, K.M. In *Amino Acids: Biosynthesis and Genetic Regulation*; Herrmann, K.M.; Somerville, R.L., Eds.; Addison-Wesley: Reading, MA, 1983, Chap. 17.
26. Frost, J.W.; Knowles, J.R. *Biochemistry* **1984**, *23*, 4465.
27. Cobbett, C.S.; Morrison, S.; Pittard, J. *J. Bacteriol.* **1984**, *157*, 303.
28. Garner, C.C.; Herrmann, K.M. *J. Biol. Chem.* **1985**, *260*, 3820.
29. Ogino, T.; Garner, C.; Markley, J.L.; Herrmann, K.M. *Proc. Natl. Acad. Sci. U.S.A.* **1982**, *79*, 5828.
30. Weaver, L.M.; Herrmann, K.M. *J. Bacteriol.* **1990**, *172*, 6581.
31. Draths, K.M.; Frost, J.W. *J. Am. Chem. Soc.* **1990**, *112*, 1657.
32. Draths, K.M.; Pompliano, D.L.; Conley, D.L.; Frost, J.W.; Berry, A.; Disbrow, G.L.; Staversky, R.J.; Lievense, J.C. *J. Am. Chem. Soc.* **1992**, *114*, 3956.
33. Gollub, E.; Zalkin, H.; Sprinson, D.B. *Methods Enzymol.* **1971**, *17A*, 349.
34. Paoletti, F.; Williams, J.F.; Horecker, B.L. *Anal. Biochem.* **1979**, *95*, 250.
35. Williams, J.F.; Blackmore, P.F.; Duke, C.C.; MacLeod, J.F. *Int. J. Biochem.* **1980**, *12*, 339.
36. Horecker, B.L. In *Comprehensive Biochemistry*; Florkin, M.; Stotz, E.H., Eds.; Elsevier: Amsterdam, 1964; Vol. 15.
37. Racker, E. In *The Enzymes*; Academic Press: New York, NY, 1961; Vol. V, Chapter 24A.
38. Blackmore, P.F.; Williams, J.F.; MacLeod, J.F. *FEBS Lett.* **1976**, *64*, 222.
39. Josephson, B.L.; Fraenkel, D.G. *J. Bacteriol.* **1974**, *118*, 1082.
40. Berthon, H.A.; Kuchel, P.W.; Nixon, P.F. *Biochemistry* **1992**, *31*, 12792.
41. Sprenger, G. A. *J. Bacteriol.* **1992**, *174*, 1707.
42. Iida, A.; Teshiba, S.; Mizobuchi, K. *J. Bacteriol.* **1993**, *175*, 5375.
43. Draths, K.M.; Frost, J.W. *J. Am. Chem. Soc.* **1990**, *112*, 9630.
44. Gross, S.R. *J. Biol. Chem.* **1958**, *233*, 1146.
45. Schweizer, M.; Case, M.E.; Dykstra, C.C.; Giles, N.H.; Kushner, S.R. *Gene* **1981**, *14*, 23.
46. Ornston, L.N.; Neidle, E. L. In *The Biology of Acinetobacter*; Towner, K.J.; Bergogne-Bérézin, E.; Fewson, C.A., Eds.; Plenum: New York, NY, 1991; pp 201-237.
47. Grant, D.J.W.; Patel, J.C. *Antonie van Leeuwenhoek* **1969**, *35*, 325.
48. Grant, D.J.W. *Antonie van Leeuwenhoek* **1970**, *36*, 161.
49. Neidle, E.L.; Ornston, L.N. *J. Bacteriol.* **1986**, *168*, 815.
50. Draths, K M.; Ward, T.L.; Frost, J.W. *J. Am. Chem. Soc.* **1992**, *114*, 9725.
51. Jensen, R.A. *Mol. Biol. Evol.* **1985**, *2*, 92.
52. Doten, R. C., Ornston, L. N. *J. Bacteriol.* **1987**, *169*, 5827.
53. Prentki, P.; Karch, F.; Iida, S.; Meyer, J. *Gene* **1981**, *14*, 289.
54. Martinez, E.; Bartolomé, B.; de la Cruz, F. *Gene* **1988**, *68*, 159.
55. Lerner, C.G.; Inouye, M. *Nucleic Acids Res.* **1990**, *18*, 4631.
56. Strøman, P.; Reinert, W.R.; Giles, N.H. *J. Biol. Chem.* **1978**, *253*, 4593.
57. Hayaishi, O.; Katagiri, M.; Rothberg, S. *J. Biol. Chem.* **1957**, *229*, 905.
58. *Chem. Mark. Rep.*, April 18, 1994.
59. Alper, J. *Science* **1993**, *260*, 1884.

RECEIVED August 4, 1994

Chapter 4

Mechanistic Study of a Catalytic Process for Carbonylation of Nitroaromatic Compounds

Developing Alternatives for Use of Phosgene

Wayne L. Gladfelter and Jerry D. Gargulak

Department of Chemistry, University of Minnesota,
207 Pleasant Street, Southeast, Minneapolis, MN 55455

Aromatic amines have been shown to be intermediates in the metal catalyzed carbonylation of nitroaromatics to aryl carbamates. Previous research established that the novel bis(methoxycarbonyl) complex, $Ru(dppe)(CO)_2[C(O)OMe]_2$, was the most abundant species present during catalysis. In this study, the complete kinetic analysis of the reaction of p-toluidine with $Ru(dppe)(CO)_2[C(O)OMe]_2$ established that the C-N bond formed by nucleophilic attack on a metal carbonyl, and that the organic product was removed from the metal by an intramolecular elimination of aryl isocyanate.

Phosgene is a highly reactive, versatile C-1 chemical that is currently used in large scale in the manufacture of isocyanates (monomers for polyurethanes) and polycarbonates. The current production of phosgene totals over 2 billion lbs/year. Accompanying its benefits, however, is its high toxicity. Notorious for its use in chemical warfare, phosgene's toxicity limits long term exposure to concentrations less than 0.1 ppm. These facts have provided the motivation for the search for alternative procedures to synthesize isocyanates.[1]

Although thermodynamically favorable, the direct carbonylation of nitroaromatics, equation 1, does not occur in the absence of a metal catalyst.

$$ArNO_2 + 3\,CO \longrightarrow 2\,CO_2 + ArNCO \qquad (1)$$

Unfortunately, lifetimes of the catalysts are so severely limited that they are of no commercial value.[1] In the 1980's, however, two groups reported that by conducting the reaction in the presence of alcohols, especially methanol, they greatly improved catalyst lifetime.[2-11] The product, a carbamate (equation 2), could be converted into the isocyanate by pyrolysis, equation 3.

$ArNO_2 + 3\,CO + MeOH \longrightarrow ArNHC(O)OMe + 2\,CO_2$ (2)
$ArNHC(O)OMe \longrightarrow ArNCO + MeOH$ (3)

The focus of our research program is to elucidate the mechanism of the high pressure synthesis of carbamates, equation 2.*(12-19)* This reaction, which involves the cleavage and formation of a total of eight bonds (not counting π-bonds), is complex. Questions as seemingly straightforward as what is the nature of the initial substrate-catalyst interaction required study. Ultimately, we hope that the fundamental understanding of the catalysis mechanism will lead to improvements in the process and even to the discovery of new reactions. This may serve to make this chemistry more attractive and may contribute to its adoption as an alternative to the phosgene-based process.

We initiated a series of *in situ* spectroscopic, kinetic and mechanistic studies of the catalytic conversion of nitroaromatics and methanol to methyl N-arylcarbamates using $Ru(dppe)(CO)_3$ where dppe = 1,2-bis(diphenylphosphino)ethane, **1**. The *in situ* spectroscopic studies were complicated by the requirement for elevated pressures during the catalysis. Using an autoclave equipped with an attenuated total reflectance (ATR) system, we monitored the changes in the infrared spectrum during the catalysis at pressures up to 100 atm.*(12, 16)* Even at low temperatures and pressures, the initial catalyst was immediately converted into a new species having metal carbonyl stretching frequencies at higher energies, **2**. Upon reaching the typical catalytic conditions a well-defined spectrum was observed that was assigned to the novel bis(methoxycarbonyl) complex, $Ru(dppe)(CO)_2$-$[C(O)OMe]_2$, **3** and **3'**.*(16)* The focus of this paper will be on the mechanism of formation of the desired C-N bond and on the mechanism of removal of the organic moiety from the metal complex. These appear to be the reactions that determine the rate of the actual catalysis.

Experimental

Standard Schlenk techniques were implemented when working with all organometallic compounds, unless otherwise stated. A nitrogen-filled Vacuum Atmospheres glove box equipped with a Dri-Train Model 40-1 inert gas purifier was employed for manipulations carried out under glove box conditions. NMR spectroscopic work was performed on a Varian-Unity 300 instrument, and infrared spectra were collected on a Mattson Polaris spectrometer. Solution gas chromatography samples were analyzed on an HP5890 series II employing a 10 m megabor FFAP-cross linked column and a flame ionization detector. All chemicals, including anhydrous methanol packed under nitrogen, were purchased from Aldrich Chemical Company, except $Ru_3(CO)_{12}$ and dppe (1,2-bis[diphenylphosphino]ethane), which were purchased from Strem Chemical. Toluene, diethyl ether and hexane were freshly distilled from benzophenone ketyl under nitrogen. Methylene chloride was distilled from calcium hydride. Compounds **1**,(20) **3**,(16,18) and $Ru(dppe)(CO)_2[C(O)NHCH(CH_3)_2]_2$ (**6**)(17) were prepared as previously described. Carbon monoxide-^{13}C in glass vessels and lecture bottles was purchased from Isotech. Throughout the paper use of **3** or **3'** refers explicity to one of the isomers of $Ru(dppe)(CO)_2[C(O)OMe]_2$. Statements referring to both isomers will be written as the formula or as **3** and **3'**.

Kinetics of the Reaction of 3 with *p*-Toluidine Between 22 and 52°C. Stock solutions of **3** and *p*-toluidine were prepared in the glove box using a 5:1 volume/volume $C_6D_6:CH_3OH$ solvent mixture. Various ratios of the solutions and pure solvent were added to 5 mm NMR tubes which were attached to 14/20 ground glass joints. The initial concentrations of **3** and *p*-toluidine of these solutions are listed in Table I. The 14/20 ground glass joints were fitted with gas adapters (Teflon valves), then the assemblies were removed from the box, the contents were frozen with liquid N_2, and the tubes were evacuated and sealed using a torch. Immediately upon warming, the tube was placed in an NMR spectrometer with the probe warmed to the desired reaction temperature. Phosphorus-31 NMR spectra were collected at specific time intervals, and the peak areas were evaluated with the integration for all species normalized to the initial concentration of **3**.

Numerical calculations for approximation of the series of differential equations described were performed using the Runge Kutta algorithm on *Mathmatica 2.0* available from Wolfram Research, Inc. The step size used in the integration was 4 minutes for analysis of the runs at 22 and 32°C and 1.5 minutes for 42 and 52°C. Comparison of the calculated and experimental data allowed the determination of the values for k_1, k_{-1}, k_2, k_{-2}, k_3, k_{-3}, k_4, k_{-4}, k_5, and $k_{5'}$. The ratio of k_4 to k_{-4} was based on the experimental value of K_4, but the specific values of these forward and reverse rates were indeterminant and input into the calculations as large numbers relative to the other rate constants.

Table I. Rate constants and activation parameters derived from numerical analysis.

T/°C	Path[a]	3[b]	ArNH$_2$[b]	k_1[c]	k_{-1}[c]	k_2[d]	k_{-2}[d]	k_3[d]	k_{-3}[d]	k_5[c]
22	5	0.034	0.41	0.0040	0.016	0.021	0.0033	0.0053	0.00092	0.00026
32	5	0.026	0.43	0.014	0.051	0.024	0.0044	0.010	0.0022	0.00092
42	5	0.030	0.36	0.040	0.14	0.019	0.0030	0.014	0.0035	0.0054
52	5	0.030	0.36	0.14	0.48	0.034	0.0031	0.027	0.0070	0.028
$\Delta H^{\ddagger}$[e]	5			21.7±0.7	20.7±0.8	2±2	-2±2	9.3±0.9	11.9±0.9	29±2
$\Delta S^{\ddagger}$[f]	5			-4±2	-4±3	-69±7	-83±5	-45±3	-40±3	16±6
22	5'	0.034	0.41	0.0040	0.016	0.024	0.0038	0.0057	0.00098	0.00041
32	5'	0.026	0.43	0.0145	0.052	0.029	0.0055	0.011	0.0023	0.0012
42	5'	0.030	0.36	0.040	0.140	0.019	0.0042	0.015	0.0023	0.0090
52	5'	0.030	0.36	0.14	0.480	0.033	0.0065	0.032	0.0020	0.035
$\Delta H^{\ddagger}$[e]	5'			21.6±0.8	20.7±0.8	0±2	2±2	10±1	4±3	29±3
$\Delta S^{\ddagger}$[f]	5'			-4±2	-5±3	-73±8	-71±6	-44±4	-68±10	14±8

SOURCE: Reprinted with permission from ref. 19. Copyright 1994.
[a] all product is assumed to form through this species. [b] Initial concentration (M). [c] Units = min^{-1}.
[d] Units = M^{-1}·min^{-1}. [e] Units = kcal/mole. [f] Units = e. u.

Reaction of *p*-Tolylisocyanate with Various Concentrations of Aryl Amine.
Five 5.0 mL solutions of *p*-toluidine in toluene and CH_3OH mixtures (5 M in CH_3OH) were prepared at the following concentrations: 9.7, 30, 45, 71 and 122 mM. To each solution in a septum-sealed vial was rapidly added 19.2 mL (0.15 mmole) of *p*-tolylisocyanate. Each vial formed visible amounts of solid N,N'-ditolylurea within 5 min at ambient temperature. After 30 min the solution in each of the vials was analyzed using gas chromatography.

Thermolysis of Ru(dppe)(CO)$_2$[C(O)NHCH(CH$_3$)$_2$]$_2$ (6). An NMR tube was attached to a 14/20 ground glass joint. Ru(dppe)(CO)$_2$[C(O)NHCH(CH$_3$)$_2$]$_2$ (5 mg) and C$_6$D$_6$ (0.5 mL) were loaded into the NMR tube in the glove box. The tube was capped with a Teflon valve, removed from the glove box and sealed under vacuum. A ^{1}H NMR spectrum was initially collected after which the tube was removed from the spectrometer, heated to 90°C for 5 min, cooled in an ice bath, and reanalyzed immediately and after standing for 1.5 h.

Results

Two isomers are formed upon reaction of aromatic amines with Ru(dppe)(CO)$_2$[C(O)OCH$_3$]$_2$. The structures, **5** and **5'**, correspond to the substitution of the two different methoxy groups. Although aryl amines do not substitute for both methoxy groups, more nucleophilic, alkyl amines such as isopropylamine rapidly form the bis(carbamoyl) complex, **6**. The latter complex has been structurally characterized and clear evidence of the hydrogen-bonding was found.*(17)* This section will summarize the results of the kinetic studies of the formation and subsequent reaction of **5** and **5'** and studies of the thermolysis of **6**.

Reaction of 3 and *p*-Toluidine Between 22 and 52°C. As described above, two isomeric carbamoyl-methoxycarbonyl complexes were observed during the reaction of 3 with ArNH$_2$ between 22 and 52°C. The concentration profile for a reaction at 22°C is displayed in Figure 1. The concentrations were based on integration of the ^{31}P NMR spectral signals of each species in solution after normalizing the integral for the total area to the initial concentration of 3. Analysis on the proton spectra for the same reaction was not possible because the high aryl amine concentration required to see appreciable 5 and 5' masked the fine details needed to determine accurately the concentration of 5 and 5'. No other signals were observed in the ^{31}P NMR spectrum except those described below.

During the initial 150 min of the reaction at 22°C, consumption of 3 and 3' occurred, and 5 reached equilibrium with 3. On a longer time scale 5' also increased in concentration. Slow, near linear formation of 1 was observed during the reaction, starting at a concentration slightly higher than zero. The non-zero initial concentration of 1 was due to a small amount of decomposition of 3 which occurred before mixing the reagents. For each run the value of 1 formed was determined and used for the initial concentration of 1 in the kinetic expression.

As the concentration of 5' increased, a singlet at 53.3 ppm in the ^{31}P NMR spectrum became visible. Its intensity was always small, typically 1/3 that of the signal due to 5'. The singlet requires that both the phosphines are equivalent, and the chemical shift is very close to those of 5 and 5'. A similar signal was observed during the reaction of 3 with isopropyl amine.*(17)* Although both structures, 8 and 9, are consistent with these data, the apparant connection between the concentration of 5' and this new species at all amine concentrations suggests that 8 (an isomer of 5' and, therefore, 5) is the more likely of the two structures. The isomerization of 3 to 3' occurs by an intramolecular "ligand-hopping" of the methoxy group from one CO to another.*(18)* The formation of 8 by the analogous event could occur directly from 5', but not from 5.

Kinetics of 3 and *p*-Toluidine Between 22 and 52°C. During the formation of 5 and 5' from 3 in the determination of the effect of CH$_3$OH and *p*-toluidine concentration on K$_2$ and K$_3$, it was observed at ambient conditions

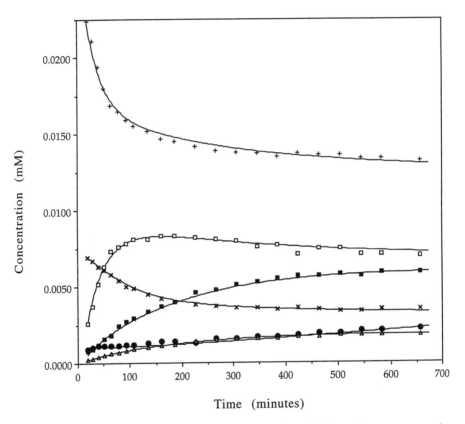

Figure 1. Experimental data and theoretical model for the progress of reaction of **3** with *p*-toluidine at 22 °C. Key: + = [**3**], x = [**3'**], □ = [**5**], ■ = [**5'**], ∆ = [**8**], • = [**1**]. Fits of similar quality were obtained for all four temperatures studied involving either **5** or **5'** as the product forming intermediate. The solution shown in this figure assumes that all product forms from **5**. (Reproduced with permission from ref. 19. Copyright 1994 American Chemical Society).

that **5** achieved equilibrium much faster than **5'**, and that their relative concentrations decreased with time resulting in the stoichiometric formation of urea and **1**. Equations 4-7 depict the reactions occurring in solution.

$$3 \underset{k_{-1}}{\overset{k_1}{\rightleftharpoons}} 3' \tag{4}$$

$$3 + ArNH_2 \underset{k_{-2}}{\overset{k_2}{\rightleftharpoons}} 5 + CH_3OH \tag{5}$$

$$3 + ArNH_2 \underset{k_{-3}}{\overset{k_3}{\rightleftharpoons}} 5' + CH_3OH \tag{6}$$

$$5' \underset{k_{-4}}{\overset{k_4}{\rightleftharpoons}} 8 \tag{7}$$

A scenario where **5** or **5'** could be responsible for product formation through a unimolecular reaction is illustrated in equations 8 and 9. The aryl isocyanate would be rapidly trapped by any free aryl amine in the solutions.

$$5 \xrightarrow{k_5} 1 + ArNCO + CH_3OH \tag{8}$$

$$5' \xrightarrow{k_{5'}} 1 + ArNCO + CH_3OH \tag{9}$$

Based on the above equilibria (equations 4-7) and the unimolecular decomposition of **5** (equation 8), the following set of rate equations and the ruthenium mass balance (equations 10-16) can be written.

$$d[3]/dt = -\{k_2 + k_3\}[ArNH_2][3] - k_1[3] + k_{-1}[3'] + k_{-2}[CH_3OH][5] + k_{-3}[CH_3OH][5'] \tag{10}$$

$$d[3']/dt = k_1[3] - k_{-1}[3'] \tag{11}$$

$$d[5]/dt = k_2[ArNH_2][3] - k_5[5] - k_{-2}[CH_3OH][5] \tag{12}$$

$$d[5']/dt = k_3[ArNH_2][3] - k_4[5'] - k_{-3}[CH_3OH][5'] + k_{-4}[8] \tag{13}$$

$$d[1]/dt = k_5[5] \tag{14}$$

$$d[8]/dt = k_4[5'] - k_{-4}[8] \tag{15}$$

$$[Ru]_{total} = [3] + [3'] + [5] + [5'] + [1] + [8] \qquad (16)$$

Numerical integration of this series of differential equations using the Runge Kutta method was successful. The mass balance for Ru species (equation 16) was implicit, and was based upon the initial concentrations. The calculated curves in Figure 1 are based on this numerical integration. Using this model, the individual rate constants were determined and are displayed in Table I. Good sensitivity was obtained for these fits, with errors on individual k values assigned as ± 10% based on the fact that deviations of this magnitude for any one constant caused considerable skewing of the model from the experimental data. The alternative procedure, in which **5'** was the productive intermediate, involved suitable changes to equations 12 - 14 and gave similar values for the rate constants (Table I). The values of K_1, K_2, and K_3 were calculated from 22 to 52°C using the data from Table I, and the values for K_4 were calculated from the ratios of **8** to **5'** obtained directly from the spectra. For increasing temperatures K_4 was 0.30, 0.41, 0.45, and 0.55, respectively.

Reactivity of 6. Thermolysis of a 14 mM solution of **6** for 5 min at 90 °C in C_6D_6 in a sealed NMR tube under vacuum resulted in a 1.8 : 1.2 : 0.2 molar ratio of diisopropyl urea (4.2 ppm, overlapping d and m, 2H; 1.1 ppm, d, 6H) : isopropyl amine (2.9 ppm, m, 1H; 0.95 ppm, d, 6H) : isopropyl isocyanate (3.0 ppm, m, 1H; 0.75 ppm, d, 6H). Upon standing for 1.5 h the residual isopropyl isocyanate was consumed, and a 2.3 : 1 ratio of diisopropyl urea to isopropyl amine was observed. The conversion of **6** to **1** was 80%, and no other metal species were observed in the ^{31}P NMR spectrum. The net reaction is shown in equation 17. This experiment was attempted at higher concentrations of **6**

$$6 \longrightarrow \{1 + RNCO + RNH_2\} \longrightarrow 1 + (RNH)_2CO \qquad (17)$$

(50 mM), but rapid urea formation masked direct observation of the isocyanate. The nonstoichiometric conversion to urea evidenced by the presence of excess amine was probably due to trace amounts of water which could react with the isopropyl isocyanate to form amine and CO_2.

Discussion

Before detailing the chemistry occurring between $Ru(dppe)(CO)_2[C(O)OMe]_2$ and amines, we will briefly review the reactivity of this unusual bis(methoxycarbonyl) complex. The two isomers, **3** and **3'**, interconvert by an intramolecular "ligand-hopping" mechanism in which a methoxy group migrates from a methoxycarbonyl to an adjacent metal carbonyl ligand.(18) The primary coordination sphere of this octahedral, d^6, Ru^{2+} complex is inert towards substitution on the timescale of all the reactions discussed in this paper. Only 10% of the carbonyls exchanged with ^{13}CO (3 atm) following three weeks of reaction at room temperature. The ester-like functional groups do undergo a transesterification upon reaction with other alcohols. Kinetic studies of the exchange of isomer 3 with CD_3OD at 21°C established that the rate of substitution of the methoxycarbonyl located trans to the phosphine was an order of magnitude higher (0.48 $M^{-1}min^{-1}$) than the rate of substitution of the group located trans to CO (0.048 $M^{-1}min^{-1}$). All OCD_3 incorporation into isomer 3' was accounted for by the rate of isomerization leading to the conclusion that 3' is inert (relative to 3) towards substitution of the methoxy groups. The mechanism of methoxy group exchange of 3 that is most consistent with the experimental data involves initial nucleophilic attack of the methanol on one of the metal carbonyl ligands.*(18)*

The reaction of $Ru(dppe)_2(CO)_2[C(O)OCH_3]_2$ with aryl amines occurs relatively rapidly at room temperature. Within 150 min the equilibrium concentration of the new monosubstituted complex, $Ru(dppe)_2(CO)_2$-$[C(O)OCH_3][C(O)NHAr]$, forms. At longer times, this complex undergoes unimolecular reaction to form $Ru(dppe)(CO)_3$, aryl isocyanate (trapped immediately by the excess amine), and methanol. This relatively simple scheme, equations 18 and 19, is complicated by the additional equilibria

$$3 + ArNH_2 \rightleftharpoons CH_3OH + 5 \tag{18}$$

$$5 \longrightarrow Ru + ArNCO + CH_3OH \tag{19}$$

involving 3. This includes the isomerization of 3 to the less reactive (towards substitution) molecule, 3', and the reaction of aryl amine with 3 to give the isomeric, monosubstituted complex 5'. This latter species can also generate the products of the reactions.

The numerical integration procedure used to determine the specific rate constants gave excellent fits to the experimental data at all temperatures studied. In addition, the rate constants produced linear Erying plots. An independent check on the procedure was possible by comparing the rate constants, k_1 and k_{-1}, for isomerization of 3 obtained by the numerical fit to the analogous rates obtained*(18)* for this isomerization as it occurs in the absence of $ArNH_2$ and methanol. Although the conditions of these two experiments differed with regard to the methanol concentration (5 vs. 0 M), the values of k_1 differed by only 7%.

We were fortunate to be able to study this system starting with samples of pure $Ru(dppe)(CO)_2[C(O)OCH_3]_2$, because the evaluation of the kinetics *during the approach to equilibrium* allowed a unique determination of the rates of most of the reactions. In the following sections we will discuss separately the results that allow the determination of the mechanisms of the N-C bond forming step and the subsequent formation of the product. In some of the possible mechanisms both N-C bond formation and product formation occur simultaneously.

C-N Bond Forming Step. Scheme 1 outlines the reasonable mechanisms that could lead to C-N bond formation. Both the reductive elimination and the migratory insertion mechanisms require the prior coordination of the aryl amide ligand. The formation of this moiety from $Ru(dppe)(CO)_2[C(O)OCH_3]_2$ would require formation of a vacant site by CO dissociation or dissociation of one end of the chelating phosphine ligand. Both of these possibilities (and as a result both the reductive elimination and migratory insertion mechanisms) can be dismissed because 1) there is no dependence of the rate on the pressure of CO (up to at least 3 atm), and 2) the rate of exchange of ^{13}CO with $Ru(dppe)(CO)_2[C(O)OCH_3]_2$ is far slower than the rate of $ArNH_2$ reaction with $Ru(dppe)(CO)_2[C(O)OCH_3]_2$. In addition, the reductive elimination mechanism would produce carbamate directly rather than the observed product, diarylurea.

Nucleophilic attack of the aryl amine on the complex could lead to displacement of the metal with direct formation of the product or to formation of the intermediate carbamoyl ligand. The former mechanism is ruled out because it would produce the wrong product, a carbamate. The formation of the carbamoyl ligand by nucleophilic attack on a metal carbonyl or on a methoxycarbonyl cannot be dismissed by any of our data. Unfortunately the site of nucleophilic attack cannot be established by our kinetics or any of our direct evidence. In a separate study of the transesterification of $Ru(dppe)(CO)_2[C(O)OCH_3]_2$, the reactivity patterns could be best explained by invoking nucleophilic attack on a metal carbonyl.*(18)* It is reasonable to expect a similar mechanism of nucleophilic attack with the aryl amine.

We conclude that the C-N bond forming step in the reaction between $Ru(dppe)(CO)_2[C(O)OCH_3]_2$ and aryl amines occurs by a nucleophilic attack, and that this attack is probably directed at a metal carbonyl. The activation parameters (Table I) calculated for this reaction exhibit large, negative entropies of activation. The magnitudes of these values are somewhat larger than those observed for the attack of nucleophiles such as methoxide*(21)* and trimethylamine-N-oxide*(22)* on $Ru(CO)_5$. This may be due to more stringent steric constraints or to a greater contribution from change in solvent ordering in the transition state for the more polar structure of $Ru(dppe)-(CO)_2[C(O)OCH_3]_2$. It is also possible that more than a single microscopic step is involved in the equilibria represented by equations 5 and 6 and that k_2 and k_{-2} (and k_3 and k_{-3}) represent a convolution of specific rate constants. Comparing this reaction to the mechanism of hydrolysis or amination of organic esters would lead one to expect a more complicated scheme which ultimately must account for all of the proton transfer steps. The increase in the rate of approach to equilibrium for equations 5 and 6 as the methanol concentration is increased is consistent with a picture where the more polar solvent mixture enhances the rate of proton transfer steps.

Product Forming Step. Scheme 2 outlines several reasonable mechanisms for removing the organic product from the catalyst. Although the reactivity of carbamoyl ligands has not been widely studied, one of the early reports of their behavior describes the elimination of isocyanate caused by deprotonation with triethylamine (equation 20).*(23)*

$$CpW(CO)_3[C(O)NHMe] + Et_3N \longrightarrow MeNCO + [CpW(CO)_3]^- + [Et_3NH]^+ \quad (20)$$

The analogous elimination from $Ru(dppe)(CO)_2[C(O)OCH_3][C(O)NHAr]$ involving $ArNH_2$ as the external base is a possible mechanism of product formation. It can, however, be dismissed because it would require the overall conversion from $Ru(dppe)(CO)_2[C(O)OCH_3]_2$ to products to exhibit second order kinetics with respect to the concentration of $ArNH_2$. This transformation is clearly first order in $ArNH_2$.

Scheme 1. Possible Mechanisms for C-N Bond Formation. $[Ru] = Ru(dppe)(CO)_2$.

a. Reductive Elimination

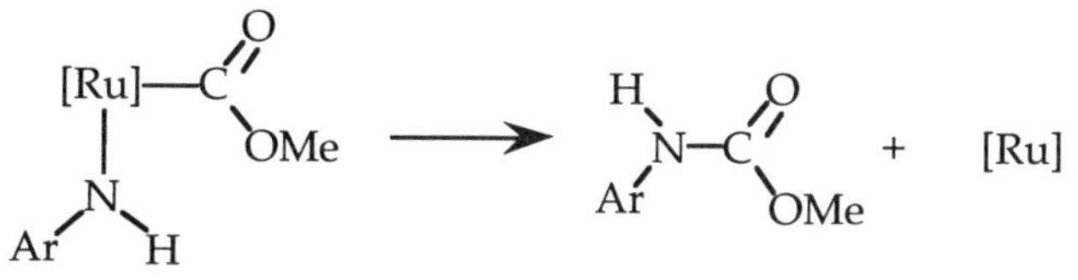

b. Migratory Insertion

c. Nucleophilic Attack on a Coordinated Ligand

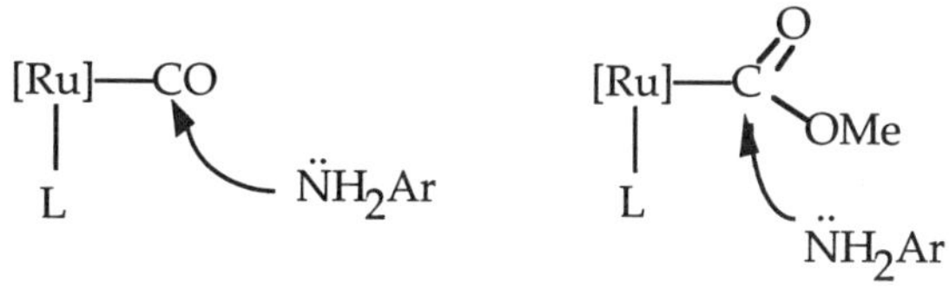

The intramolecular elimination of aryl isocyanate is consistent with the observed kinetics, and the structure of $Ru(dppe)(CO)_2[C(O)OCH_3]$-$[C(O)NHAr]$ is ideally posed to undergo this reaction. Although this mixed methoxycarbonyl-carbamoyl complex could not be isolated, its ^1H NMR spectrum exhibited a downfield shift of the amido hydrogen characteristic of a hydrogen bond to the oxygen of the methoxycarbonyl ligand. The x-ray crystallographic study of the bis(carbamoyl) complex, $Ru(dppe)(CO)_2$-

[C(O)NH-*i*-Pr]$_2$, identified the details of this hydrogen-bonded six-membered ring.(17) This complex eliminates isopropyl isocyanate and isopropyl amine upon thermolysis providing excellent precedent for the reaction of **5** and **5'**. A concerted rearrangement illustrated in Scheme 2 allows the expulsion of aryl isocyanate and leaves a hydroxy-methoxy carbene which would readily lose methanol and form Ru(dppe)(CO)$_3$. The slightly positive values for activation entropy can be understood because the ground-state structures of **5** and **5'** are already organized in a constrained arrangement that can lead to the products.

The last possible mechanism for removal of the organic product shown in Scheme 2 is referred to as a cycloreversion. Regardless of whether this is proposed to occur from **5** or **5'** or even a bis(carbamoyl) complex, **9**, direct experimental evidence rules out this mechanism. Cycloreversion from **5** or **5'** would lead directly to carbamate, the wrong product, and cycloreversion from **9** would require second order kinetics with respect to the concentration of aryl amine.

Scheme 2. Possible Product Forming Steps. [Ru] = Ru(dppe)(CO)$_2$.

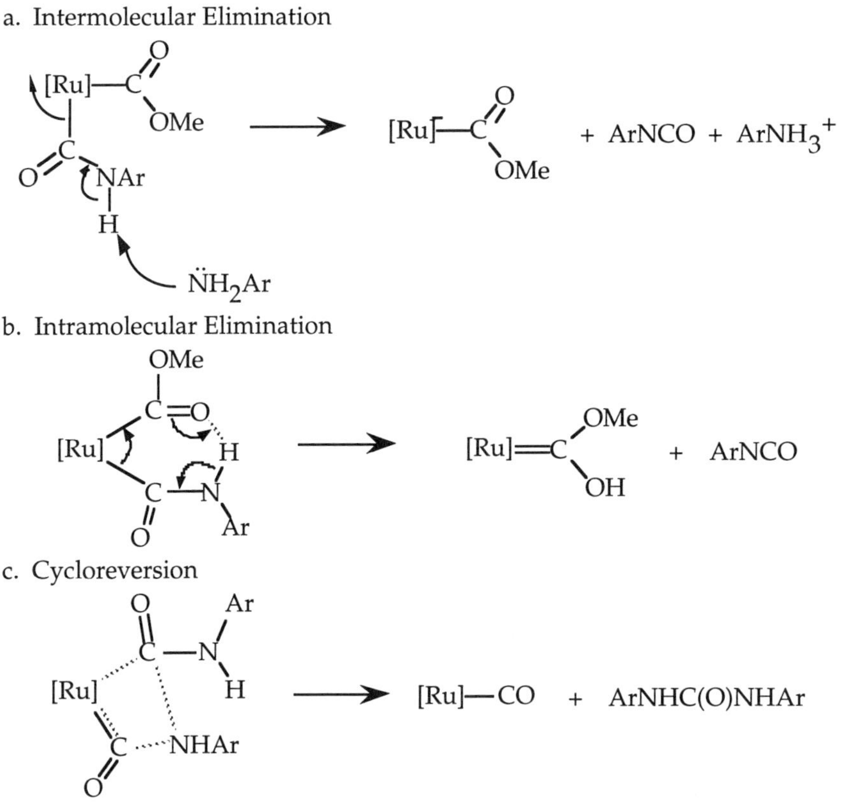

a. Intermolecular Elimination

b. Intramolecular Elimination

c. Cycloreversion

We conclude that the intramolecular elimination of aryl isocyanate is the mechanism which releases the organic product from the catalyst. In all of the stoichiometric reactions this isocyanate is immediately trapped by $ArNH_2$ to form the observed product, diarylurea. When the reaction was conducted under catalytic conditions either diarylurea or the carbamate was observed depending upon the conditions. Even in those runs that produced carbamates, urea was still proposed to be the intermediate.(5)

The results discussed in this and the preceding section can be effectively summarized on the reaction coordinate diagram shown in Figure 2, where the values of free energy were calculated using a temperature of 22°C and the standard state concentration of 1 M. The equilibria between **3** and $ArNH_2$ are represented with one barrier even though these reactions involve several microscopic proton transfer steps in addition to the nucleophilic attack and cleavage of the C-O bond. Recall that **5** and **5'** are isomers of each other so that their similar reactivity is not surprising. It is worth reiterating that in the determination of the rate constants for the conversion of **5** to products, we ignored the reaction of **5'** to products. Likewise the rate constant involving **5'** was obtained by shutting down product formation from **5**. Both of the mechanisms gave equally good fits to the experimental results, and we would expect to be able to fit the data with any linear combination (e. g. 30% of the product forms from **5** and 70% from **5'**) between the two extreme cases. We believe that the similarities in the reaction coordinate diagram, despite the fact that the numbers were determined for the extreme cases, indicate that both isomers are contributing to product formation.

Summary

In this paper we have established that the mechanism of C-N bond formation involves nucleophilic attack of $ArNH_2$ on a metal carbonyl ligand of $Ru(dppe)(CO)_2[C(O)OMe]_2$. The carbamoyl intermediate undergoes a unimolecular elimination of aryl isocyanate, which is rapidly trapped to give the products of the catalysis. Figure 3 incorporates the details of this study into a summary catalytic cycle. Unanswered questions about the catalytic cycle remain, especially regarding the early stages. Work is continuing that focuses on the mechanism of the activation of the starting nitroaromatic and the mechanism of cleavage of the N-O bonds.

Impact on the Goal of Replacing Phosgene

Knowledge of the mechanism of the rate determining step allows us to predict the effect of changing the structure of the metal complex and other parameters of the catalysis. For instance, making the phosphorous ligand smaller in cone angle and less electron donating should increase the activity of the catalyst. This change has yet to be tested.

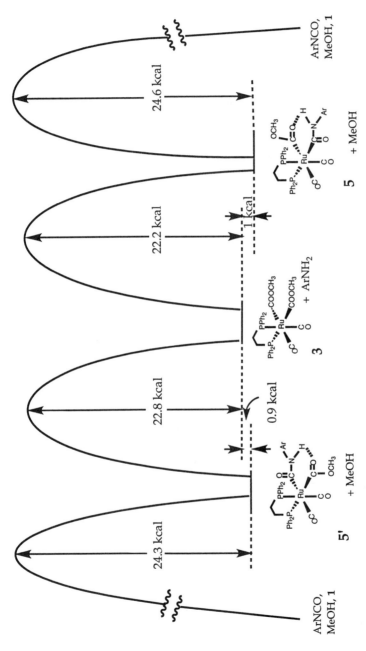

Figure 2. Energy level diagram for the transformations of 3 to products. The equilibrium between 3 and 3' is not shown. The values are calculated for 22°C, and the standard states of the components are 1 M. (Reproduced with permission from ref. 19. Copyright 1994 American Chemical Society).

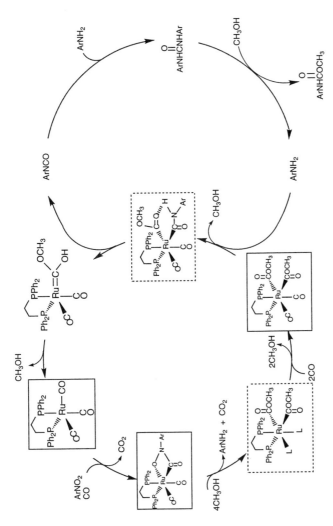

Figure 3. Summary of the mechanism of the overall catalytic cycle. The ruthenium complexes within solid boxes have been isolated and fully characterized. Those within dashed boxes have been observed, but neither fully characterized nor isolated. The identity of the two ligands, labeled L, in the precursor to **3** remains obscure. The hydroxymethoxycarbene has not been observed. (Reproduced with permission from ref. 19. Copyright 1994 American Chemical Society).

The most important single use of phosgene is for the synthesis of toluene diisocyanate (TDI), and any significant impact on the volume of phosgene currently consumed must involve an alternative preparation of this compound. One of the shortcomings of all of the catalytic systems for nitroaromatic carbonylations, including Ru(dppe)(CO)$_3$, is the difficulty they have in carbonylating dinitroaromatic substrates, such as 2,4-dinitrotoluene (the precursor to TDI). Consideration of the rate determining step in the Ru(dppe)(CO)$_3$ catalytic system reveals some insight into the cause of this problem. The substrate would have to proceed around the cycle two times for complete carbonylation. After the first nitro group is reduced, the resulting nitroaminotoluene would have significantly reduced nucleophilicity and thus may proceed sluggishly through the critical C-N bond forming stage. This would allow the catalyst sufficient time to cycle back through alternative paths to Ru(dppe)(CO)$_3$ which would lower both the selectivity and yield. Designing catalysts that can tolerate the higher degree of functionality required for wide application remains an important goal.

At this time cost estimates suggest the catalytic routes are close to the phosgene-based methods for mononitroaromatic substrates. Whether this technology is ultimately adopted as an alternative, phosgene-free route to carbamates and isocyanates will probably be determined by changes in regulations governing the use of phosgene. These factors are difficult and beyond our expertise to predict. What we can conclude is that the catalytic chemistry is selective and gives high yields, is reproducible, and is well-behaved even from the mechanistic perspective.

Acknowledgments

This research was supported by a grant from the National Science Foundation (CHE-9223433). Fellowship support from Hercules (for J. D. G.) is gratefully acknowledged as are the valuable discussions with John Grate and David Hamm.

Literature Cited

1. Cenini, S.; Pizzotti, M. Crotti, C. In *Aspects of Homogeneous Catalysis*; R. Ugo, Ed.; Reidel: Dordrecht, 1988; Vol. 6; pp 97 - 198.
2. Grate, J. H.; Hamm, D. R. Valentine, D. H. U. S. Patent 4600793, 1986.
3. Grate, J. H.; Hamm, D. R. Valentine, D. H. U. S. Patent 4603216, 1986.
4. Grate, J. H.; Hamm, D. R. Valentine, D. H. U. S. Patent 4629804, 1986.
5. Grate, J. H.; Hamm, D. R. Valentine, D. H. U. S. Patent 4705883, 1987.
6. Cenini, S.; Pizzotti, M.; Crotti, C.; Porta, F. La Monica, G. *J. Chem. Soc., Chem. Commun.* **1984,** 1286.
7. Cenini, S.; Crotti, C.; Pizzotti, M. Porta, F. *J. Org. Chem.* **1988,** *53,* 1243.
8. Cenini, S.; Pizzotti, M.; Crotti, C.; Ragaini, F. Porta, F. *J. Mol. Catal.* **1988,** *49,* 59.

9. Cenini, S.; Ragaini, F.; Pizzotti, M.; Porta, F.; Mestroni, G. Alessio, E. *J. Mol. Catal.* **1991**, *64*, 179.
10. Ragaini, F.; Cenini, S. Demartin, F. *J. Chem. Soc., Chem. Commun.* **1992**, 1467.
11. Bassoli, A.; Rindone, B.; Tollari, S.; Cenini, S. Crotti, C. *J. Mol. Catal.* **1990**, *60*, 155.
12. Kunin, A. J.; Noirot, M. D. Gladfelter, W. L. *J. Am. Chem. Soc.* **1989**, *111*, 2739.
13. Gargulak, J. D.; Noirot, M. D. Gladfelter, W. L. *J. Am. Chem. Soc.* **1991**, *113*, 1054.
14. Gargulak, J. D.; Hoffman, R. D. Gladfelter, W. L. *J. Mol. Catal.* **1991**, *68*, 289.
15. Sherlock, S. J.; Boyd, D. C.; Moasser, B. Gladfelter, W. L. *Inorg. Chem.* **1991**, *30*, 3626.
16. Gargulak, J. D.; Berry, A. J.; Noirot, M. D. Gladfelter, W. L. *J. Am. Chem. Soc.* **1992**, *114*, 8933-8945.
17. Gargulak, J. D. Gladfelter, W. L. *Inorg. Chem.* **1994**, *33*, 253 - 257.
18. Gargulak, J. D. Gladfelter, W. L. *Organometallics* **1994**, 698 - 705.
19. Gargulak, J. D. Gladfelter, W. L. *J. Am. Chem. Soc.* **1994**, in press.
20. Sanchez-Delgado, R. A.; Bradley, J. S. Wilkinson, G. *J. Chem. Soc., Dalton Trans.* **1976**, 399-404.
21. Trautman, R. J.; Gross, D. C. Ford, P. C. *J. Am. Chem. Soc.* **1985**, *107*, 2355-2362.
22. Shen, J.-K.; Gao, Y.-C. Basolo, F. *Organometallics* **1989**, *8*, 2144-2147.
23. Jetz, W. Angelici, R. J. *J. Am. Chem. Soc.* **1972**, *94*, 3799.

RECEIVED September 16, 1994

Chapter 5

Preparative Reactions Using Visible Light
High Yields from Pseudoelectrochemical Transformation

Gary A. Epling and Qingxi Wang

**Department of Chemistry, University of Connecticut, U-60,
215 Glenbrook Road, Storrs, CT 06269**

Methods to prevent pollution from toxic heavy metals have been explored through development of nontoxic catalytic alternatives for oxidation of dithianes, oxathianes, and benzyl ethers. Inexpensive spotlights were utilized to induce high-yield preparative scale photoreactions. In this manner, dye-sensitized irradiation of dithianes led to cleavage, generating a dithiol and a carbonyl compound, while oxathiane irradiation led to a carbonyl compound and a hydroxythiol. Trimethoxybenzyl ether irradiation produced an alcohol and an aromatic aldehyde. A single electron transfer mechanism seems responsible for initiating the bond cleavages that mimic electrochemical oxidation of these substrates. An advantage over electrochemistry, however, is that the dye-sensitized transformations are more conveniently scaled up. In this study multi-gram reactions have been performed without difficulty.

In recent years the costs of decades of environmental neglect have begun to be truly appreciated by government, industry, and the general public. Adversarial debates have begun to give way to increasing agreement on sustainable goals for technological improvements that will lead to technologies for cleaning up pollution, design of products that are less polluting over their useful life, and changing the way products are manufactured so that less pollution is created during the manufacturing process. The Environmental Protection Agency developed one of the earliest well-articulated approaches to a safer environment through the concept of pollution prevention. The clear-cut financial advantages of pollution *prevention* over pollution *remediation* technologies were the underlying motivation for the EPA's 33/50 Program which targeted 17 chemicals for examination, requesting that companies voluntarily initiate efforts to effect a 50% reduction of the release of these chemicals by 1995.

The 17 target chemicals of the 33/50 program were chosen using three criteria: they pose environmental and health concerns, they are high-volume industrial chemicals, and their release can be reduced through pollution prevention. The targets thus selected were:

Benzene	Methyl Ethyl Ketone
Cadmium Compounds	Methyl Isobutyl Ketone
Carbon Tetrachloride	Methylene Chloride
Chloroform	Nickel and Compounds
Chromium and Compounds	Tetrachloroethylene
Cyanides	Toluene
Lead and Compounds	Trichloroethane
Mercury and Compounds	Trichloroethylene
	Xylenes

A host of useful organic compounds are prepared using synthetic procedures that rely upon the use of these solvents or reagents. Many oxidation reactions, for example, commonly utilize lead, chromium, manganese, mercury, thallium, or cerium compounds as oxidants. Consequently, the beneficial effects that result from preparation of a useful compound are partially offset by the concomitant adverse considerations from use of potential environmental pollutants. Environmentally-conscious behavior would include detoxification or safe disposal of such pollutants. There exists no convenient technology for inexpensive quantitative recovery and recycle of many reagents, like the ones described above. Consequently, a superior long-range strategy would be to develop alternative methods for achieving desired synthetic transformations using substitute reagents that are nontoxic.

Our research focuses on development of synthetically-useful reactions as replacements for existing methodology that uses undesirable toxic reagents. In particular, we have initially focused on oxidation reactions because of the high toxicity of the reagents and the widespread application of such transformations. Our approach, in essence, is to utilize visible light as the "reagent" which provides the thermodynamic driving force for a synthetic transformation. In this regard, the descriptor "green chemistry" is more than an accident, because the underlying mechanism that we rely upon for the key electron transfer step has a direct analogy to the photosynthetic mechanism utilized by green plants in the selective oxidation of water and reduction of carbon dioxide. A second goal of our research is to focus attention on synthetic reactions for which alternative methods of reaction might expand upon the synthetic options presently available and provide improvements in yield or selectivity.

We chose to use visible light rather than ultraviolet light for several reasons. The physical generation of ultraviolet light requires modestly expensive light sources, and even a limited scaleup of the reaction requires a fairly large capital investment. Further, the high energy of ultraviolet light is readily absorbed by many chromophores, sometimes leading to a host of reactions when complex substrates are irradiated. Quantitative transformations utilizing ultraviolet light are so rare they seem almost fortuitous. Finally, the lower energy of visible light leads to a greater likelihood of a higher selectivity of reaction, with resulting higher yields and minimal byproduct formation.

Intense sources of visible light are readily available and inexpensive. In the early stages of our work we have primarily utilized inexpensive spotlights of the type readily available at hardware stores and discount stores. These lights have been used to perform reactions of up to 20 g of substrates, sometimes in only 12 to 24 hours of illumination. With inexpensive sources that are more powerful, or by using a solar collector for sunlight, it ought to be feasible to perform reactions on the kilogram scale, particularly if improvements of efficiency result from additional research along avenues that we envisage.

To harness the energy of visible light into a productive oxidation reaction we need a dye to absorb the light and then participate in an efficient electron transfer reaction. Further, the dye should function catalytically rather than stoichiometrically. The dye should be inexpensive and nontoxic. A significant portion of our work has focused on the selection of appropriate dye catalysts. Several good candidates have emerged, some of which have sufficiently low toxicity that they are used as food colorants, cosmetics, or dyes in medicines.

We have chosen three oxidation reactions for initial study—cleavage of dithianes to aldehydes and ketones, cleavage of benzyl ethers to alcohols, and cleavage of oxathianes to the corresponding carbonyl compound and the thioalcohol. These reactions were selected for several reasons. First, the common methods for accomplishing the indicated transformations require toxic reagents. Second, there are many compounds for which existing synthetic methods either fail to accomplish the desired transformation or proceed in low yield. Finally, theoretical considerations suggested that the desired oxidations should be possible using the dye-promoted methodology we proposed. We are pleased to report that these transformations have been accomplished cleanly, and that our method appears to offer a practical alternative to the use of toxic heavy metals for these purposes.

Chemical Transformations Chosen for Examination. Dithio acetals and ketals are useful protecting groups for carbonyls, and are widely used in organic synthesis (*1*). As a carbonyl protecting group they complement the role of acetals and ketals. Dithioacetals have an acidic hydrogen which can be deprotonated to transform the masked carbonyl into a nucleophile. Finally, the parent 1,3-dithiane, by such methodology, can be used as a reagent for a one-carbon homologation. Nevertheless, their utility has been somewhat curtailed by the sometimes problematic deprotection step and the absence of appropriate methods that proceed under mild conditions. We sought to develop a procedure for deprotection of dithianes and dithiolanes that would proceed under neutral conditions without the use of heavy metals or chemical oxidants.

1,3-Oxathianes are versatile protecting groups for carbonyls which have utility like that of dithianes (*2*). Their deprotection is similarly accomplished, most often relying on oxidative methods which utilize heavy metals. Probably the most common methods involve mercuric chloride in acetic acid, mercuric chloride with alkaline ethanolic water, or silver nitrate with N-chlorosuccinimide. One of the interesting new applications of the oxathiane protecting group is stereo-controlled "asymmetric synthesis," pioneered by the elegant work of Eliel (*3*). Unfortunately, synthesis of the requisite "chiral auxiliary" from pulegone is not straightforward, and cleavage of the chiral protecting group has also been somewhat troublesome. Further, the best method for cleavage (N-chlorosuccinimide/silver nitrate) does not allow the direct recovery of the chiral thioalcohol, but an oxidized sultine, which must be reduced subsequently in order to regenerate the precious chiral auxiliary (*4*). We have focused attention on Eliel's chiral oxathianes to determine whether our pseudoelectrochemical methodology would provide a cleavage method that might proceed under neutral conditions and allow direct recovery of the chiral auxiliary.

The benzyl ether protecting group is a common way to protect an alcohol during synthetic transformations of a complex molecule containing the hydroxyl functionality. However, the usual ways to remove this blocking group (i.e., catalytic hydrogenation or alkali metal reduction) involve conditions which sometimes may not be applied to molecules containing easily reduced functional groups (*5*). Because of the precedented cleavage of this group using electrochemical methods (*6*), we believed a light-driven electron transfer reaction could be used for cleavage of this group by what would formally be considered an oxidative pathway, providing an alternative to a cleavage using strong reductants.

Results and Discussion

The essence of our approach was to determine whether bond cleavages leading to the desired transformations might be initiated by photoinduced electron transfer using visible light. To this aim we selected dyes that might trigger the desired electron transfer process in high efficiency with a minimum rate of back electron transfer and a maximum rate of separation of the resulting radical ions. In this manner a rapid net chemical reaction of the starting material would be achieved. The considerations regarding appropriate dyes, and a systematic examination of structure/activity relationships will be the subject of later communications. However, we presently are pleased to report that several promising candidates have emerged from our study, and these have served well in producing the desired transformations described above.

Cleavage of the Dithiane Protecting Group. We prepared dithianes and dithiolanes of a variety of compounds using minor modifications of existing methods (7). The target molecules were chosen to provide a variety of types of materials, including simple aliphatic and aromatic compounds, but also including some compounds which contained multiple functionality. Acetonitrile–water solutions of these dithio derivatives *were* efficiently hydrolyzed when an ordinary tungsten spotlight was used as the light source; methylene green, methylene blue, rose bengal, erythrosin, and eosin B are dyes which were used as photocatalysts.

$$\underset{n = 1, 0}{\overset{R_1 \; S}{\underset{R_2 \; S}{\bigtimes}}(CH_2)_n} \quad \xrightarrow[\text{CH}_3\text{CN/H}_2\text{O} \atop 1:1]{\text{Methylene Green} \atop \text{visible } h\nu} \quad \underset{R_1 \quad R_2}{\overset{O}{\|}} \; + \; \underset{HS}{\overset{HS}{\diagdown}}(CH_2)_n$$

With a wide variety of dithianes (Table I) we found the dye-promoted photocleavage led smoothly to formation of the desired carbonyl compound, along with the 1,3-propanedithiol which was the original precursor to the starting dithiane. The presence of either an aromatic ring or a heteroaromatic ring did not cause difficulties. Similarly, double bonds (either conjugated or isolated) were not oxidized. Dithiane cleavage preferentially occured in the presence of an amino group. Even a ketal group was retained after cleavage of a dithiane by this methodology.

Yields reported in Table I are isolated yields of carbonyl compounds that were spectroscopically pure (NMR and GC/MS). Typically, byproducts were not formed in significant amounts, and purification was straightforward. Extraction and chromatography over a few inches of silica gel gave products of excellent purity.

Our exploratory studies were successfully scaled up to multigram quantities; 3- to 15-g quantities of dithio compounds were reacted in the examples shown in Table I. In our largest scale reactions we used 4-L round-bottomed flasks, mixtures of dyes, and two spotlights in order to achieve complete transformation in the minimum time. The progress of reaction was monitored by TLC or GC, and illumination continued until the starting material was completely consumed.

The reactions summarized in Table I were performed as solutions in acetonitrile/water. Our main considerations were good solubility of both the dye and the dithiane, while we also desired a high level of polarity for stabilization of developing ion pairs. We have recently begun to explore alcohol/water solvents, with preliminary indications they would be satisfactory substitutes for acetonitrile.

Table I. Isolated Yields of Aldehydes or Ketones from
Dye-Promoted Photocleavage of Dithio Compounds

Entry Number	Substrate	Product	Isolated Yield
1			97%
2			94%
3			95%
4			91%
5			86%
6			91%
7			95%

We routinely obtained high yields of isolate, purified aldehydes or ketones (Table I). In some cases irradiation under oxygen led to somewhat lower yields due to formation of small amounts of oxidized products, particularly for cleavages of precursors to aldehydes. However, irradiation under nitrogen (though proceeding more slowly) led to high yields even with these compounds.

Several experimental observations led us to believe the mechanism of deprotection involves electron transfer as the key step rather than a dye-sensitization of singlet oxygen:

(1) The transformations could be induced even in an nitrogen atmosphere, where they proceeded more cleanly in some cases than under oxygen.

(2) Irradiation of a dithiane under the same conditions except *omitting* the dye sensitizer led to *no* detectable reaction of the dithiane.

(3) Stirring the dithiane under the same conditions (with the dye present) but *in the dark* led to no reaction of the dithiane.

(4) The role of the dye was catalytic. Reaction of a 30 to 100–fold molar excess of the dithio compound proceeded without any observable loss of the dye, as verified by UV/VIS spectroscopy.

(5) The formation of the carbonyl product was accompanied by the formation of either 1,2-ethanedithiol or 1,3-propanedithiol (from dithiolanes or dithianes, respectively).

(6) The deprotection reaction proceeded faster in the presence of inorganic salts, suggesting an acceleration of the efficiency of separation of the radical ions formed initially.

These observations lead us to suggest the reaction mechanism for the transformations outlined as Scheme I.

Scheme I. Proposed Mechanism for Dye-Promoted Photocleavage of Dithio Compounds to Aldehydes and Ketones

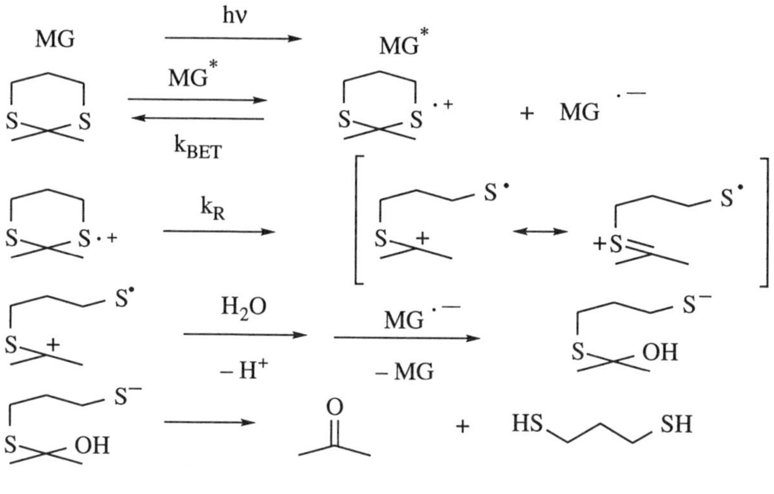

The first step is consistent both with the critical role played by the dye and the requirement for illumination of the reaction vessel. The next two steps of Scheme I are similar to the pathway proposed by Steckhan to explain the electrocatalytic cleavage of dithianes (8). Other elements of our scheme differ since we need to explain other observations. First, the role of the dye is catalytic, being temporarily reduced in the second step to the radical anion, but regenerated by back electron transfer ("k_{BET}") to the ring-opened intermediate in the fourth step. This back electron transfer competes with the bond-cleaving step which leads to product ("k_R"). Our mechanistic scheme is consistent with the overall *hydrolytic* nature of the cleavage; neither catalyst nor any part of the dithiane is oxidized or reduced. This feature distinguishes our route from previously reported photooxidations or electrochemical cleavage, as these paths lead to the formation of an oxidized disulfide. Our proposed formation of a hemithioacetal seems plausible since this intermediate can be hydrolyzed readily by acid or base to the carbonyl products observed.

Cleavage of the Oxathiane Protecting Group. Examination of Scheme I suggested a second group of compounds that might be amenable to dye-induced cleavage *via* a photolytic electron transfer mechanism—the family of oxathianes. Since the key step that led to bond cleavage was electron transfer from a sulfur, the major role of the second sulfur was to facilitate electron transfer by virtue of the lower oxidation potential of the dithio derivatives. The remaining bond cleavage (the carbon-oxygen bond) would be from a hemiacetal or hemiketal if an analogous mechanism were to occur. With this overall strategy in mind, we focused attention on the synthetic advantages that such a cleavage would offer in the deprotection of oxathianes.

We noted in the introduction that oxathianes have been used as chiral auxiliaries in the synthesis of optically active compounds of potential synthetic and medicinal importance. Eliel's 1,3-oxathiane derived from pulegone is probably the most noteworthy example. Eliel's asymmetric synthesis proceeded in three steps. The first step was reaction of the lithium salt of the chiral 1,3-oxathiane. The second step involved the reaction of the ketone obtained from the oxidation of carbinol with a Grignard reagent or metal hydride. The last step of Eliel's scheme for asymmetric synthesis was the cleavage of the oxathiane to afford optically active products and recovery of the chiral adjuvant. The recovery of this chiral auxiliary is very important in a practical asymmetric synthesis. Unfortunately, attempts to recover this oxathiol directly in good yield have been unsuccessful. Eliel has reported an efficient method for the cleavage, though his method needs toxic reagents (silver salts). Further the necessary hydroxythiol moiety was not formed directly, but a sultine resulted, which was then reduced by lithium aluminum hydride to the recovered chiral hydroxythiol.

We focused attention on several oxathianes which are important in Eliel's methodology. Like the dithianes, we found that they could be cleanly deprotected to give the desired products (Table II). As one might have anticipated, however, the deprotection proceeded somewhat more slowly than with the dithianes, though in other regards the reaction proceeded similarly. Table II shows our successful cleavage of an oxathiol group flanked either by an alcohol or a carbonyl. Like with the dithiane family of Table I, the masked carbonyl was recovered in high yield. Of great importance were the similarly high yields for direct recovery of Eliel's chiral precursor. As we had hoped, this direct recovery obviates the need for subsequent synthetic manipulation of this key reagent, while simultaneously avoiding the potential pollution that would result from the use of the heavy metal oxidant in Eliel's methodology.

As in the dithiane family, the yields of Table II are isolated yields of spectroscopically pure products. It was only necessary to follow an extractive workup with flash chromatography to achieve the desired level of purity of products.

Table II. Isolated Yields of Aldehydes and Hydroxythiol from Deprotection of Oxathiane Derivatives

Substrates	Products (Isolated Yields)		
[oxathiane with CH(Ph)OH]	[menthyl-OH, SH] (90%)	+	PhCH(OH)CHO (90%)
[oxathiane with C(Ph)=O]	[menthyl-OH, SH] (92%)	+	PhC(O)CHO (90%)
[oxathiane with C(Ph)(CH₃)OH]	[menthyl-OH, SH] (91%)	+	PhC(CH₃)(OH)CHO (91%)

Our observation of high-yield photocleavage of the chiral oxathianes suggests both a potential use for the photocleavage methodology, and also increases the potential value of Eliel's strategy for asymmetric synthesis.

Cleavage of the Benzyl Ether Protecting Group. If a dye-promoted photochemical electron transfer cleavage of a benzyl ether group were to occur it would be expected that the lower oxidation potential afforded by methoxy substitution would accelerate the rate of photocleavage. Accordingly, 3,4,5-trimethoxy-substituted benzyl ethers were prepared for initial study. The requisite benzyl ethers were prepared by a Williamson ether synthesis involving the sodium salt of the alcohol to be protected and the 3,4,5-trimethoxybenzyl chloride. The required benzyl chloride was obtained from the corresponding alcohol by treatment with thionyl chloride. Our general procedure, as before, involved irradiation of acetonitrile/water (1:1) solutions of 0.10 to 20.0 g of the benzyl ethers in the presence of 0.10 to 1.0 mM of the sensitizing dye. A spotlight was used as the irradiation source, and nitrogen was bubbled through the solution during some experiments, which resulted in cleaner transformations, though the required reaction times were longer. Under these conditions the proposed cleavage reaction ensued, and the deprotected alcohol was recovered by chromatography. Alternatively, a bisulfite extraction successfully removed the aldehyde, and high yields of the deprotected alcohol in very high purity were obtained without chromatography. Substituted benzyl ethers that were successfully cleaved in this manner are shown in Table III, along with the isolated yield of purified alcohol product.

Table III. Isolated Yields of Alcohols from Dye-promoted Photocleavage of Trimethoxy Benzyl Ethers

Substrate	Isolated Yield
3,4,5-trimethoxybenzyl N-acetylpiperidinyl ether	90%
3,4,5-trimethoxybenzyl 2-pyridylmethyl ether	96%
3,4,5-trimethoxybenzyl (CH$_2$)$_5$-tetrahydropyranyl ether	91%
3,4,5-trimethoxybenzyl piperidinyl ether	100%
3,4,5-trimethoxybenzyl pinenyl ethyl ether	91%
3,4,5-trimethoxybenzyl bornyl ether	87%

In the examples shown, a variety of functionalities were found to be amenable to this method of cleavage. Some compounds were irradiated under oxygen, which resulted in faster reaction in most cases, though in other cases it led to lower yields because of side reactions. Small amounts of competing cleavage of amines (9) were observed in some cases, though conversion to an amide, irradiation in dilute acid, or irradiation under oxygen led to clean reactions without observable amounts of cleavage of the amine.

In cases where a large quantity of compound was to be reacted, a mixture of dyes was used to harvest as fully as possible the full spectrum of light emitted by the spotlight. In this manner 20 g of the trimethoxybenzyl ether of nopol was cleaved in 89% isolated yield with only 12 hours of irradiation. Having achieved success in this larger scale reaction, we have demonstrated an apparently practical laboratory-scale procedure for removal of the trimethoxybenzyl ether protecting group. Our procedure was capable of being scaled up to multigram quantities, and used a simple experimental protocol. It appears that further study of both the generality of this reaction and methods for improvement of reaction efficiency are warranted.

Future Studies. The examples we have examined show great promise for further development of the protocol of visible light-induced photochemical transformations. We plan to proceed further to study how we can improve the overall "rate" of the transformation as well as expand our study of the types of synthetic transformations that may be feasible. Alternative solvents, are under study, in the event that recycle of the acetonitrile is not viewed as practical on a commercial scale. The success of these future studies seems to offer a great opportunity for pollution prevention technologies for the fine chemical industry. Our "pseudoelectrochemical" transformations embody a clearcut strategy for avoidance of heavy metal pollution by their direct replacement as chemical oxidants. The replacements (dye catalysts and light) are particularly attractive because of their low cost and low toxicity, but particularly since the organic dyes are only needed in catalytic amounts. We are very hopeful that future developments of this concept will lead to a significant reduction in the use of toxic metal oxidants.

Experimental

Preparation of Dithianes. Samples of 3.0-4.0 g of 1,3-propanedithiol (1-1.5 mol relative to the carbonyl compound), 2-5 mL of fresh boron trifluoride etherate (1-1.5 mol relative to the carbonyl compound), and 50 ml of dry THF (treated with sodium) were added to a 250 ml three-necked round bottom flask. After cooling the solution to 0^0C in an ice bath, a solution of 4.0 g of ketone or aldehyde in 50 mL of dry THF was added dropwise with constant stirring, under nitrogen. The ice bath was then removed and the reaction mixture stirred overnight to insure complete reaction. The reaction mixture was poured into a slurry of crushed ice and water containing 5% ammonium chloride, then extracted with ethyl ether. The combined organic layers were washed with 5% sodium carbonate aqueous solution, dried, and concentrated to give the crude product. Purification by column chromatography afforded spectroscopically pure dithianes.

General Procedure for Photosensitized Deprotection of Dithianes Using Visible Light. A solution of ca. 3.0 g (15 mmol) dithiane, 182 mg (0.5 mmol) methylene green (or an equivalent weight of rose bengal, erythrosin, methylene blue, or eosin B), and 440 mg (2 mmol) magnesium perchlorate in 500 mL of acetonitrile:water (1:1) was placed in a 500 mL round bottom flask equipped with a reflux condenser. A polyethylene tube was inserted into the solution and used to purge the solution with nitrogen or oxygen during the reaction. The flask was illuminated with a 120W

GE Miser or 150W Sylvania indoor spotlight until reaction was complete (generally 4–24 hours). After TLC showed that the reaction was complete, the product was isolated by adding 200 mL of 1% hydrochloric acid, saturating the solution with sodium chloride, and extracting with chloroform. The chloroform extracts were washed, dried with anhydrous sodium sulfate, and concentrated to give the crude product. Purification by flash chromatography gave spectroscopically pure aldehyde or ketone. Purity and identity were confirmed by NMR and GC/MS.

Preparation of and Transformations of Oxathianes derived from Pulegone. These materials, shown in Table II, were prepared using modifications of the synthetic procedures of Eliel (*10*).

General Procedure for Photosensitized Deprotection of Oxathianes Using Visible Light. A solution of 0.1-1 g oxathiane and 182 mg (0.1 mmol) methylene green (or an equivalent weight of rose bengal, erythrosin, methylene blue, or eosin B) in 100 mL of acetonitrile:water (1:1) was placed in a 200 mL pyrex tube. A polyethylene tube was inserted into the solution and used to purge the solution with oxygen during the reaction. The solution was illuminated for 6–12 hr with a 120W GE Miser or 150W Sylvania indoor spotlight. After TLC showed the reaction was complete, the reaction mixture was saturated with sodium chloride and extracted with chloroform. The combined organic layers were washed with saturated brine, dried, and concentrated to give the unprotected aldehyde or ketone. Finally, purification by flash chromatography gave spectroscopically pure product. The structure was confirmed by NMR and MS.

Preparation of Benzyl Chlorides. A 5.0 g sample (0.015-0.025 mol) of methoxy substituted benzyl alcohol in 100 mL benzene was added to a 250 mL round bottom flask. With constant stirring, 3.0 ml (1-1.5 mol relative to the benzyl alcohol) thionyl chloride was added dropwise over a period of 30 min. The reaction mixture was refluxed, and then transferred into a 500 ml beaker containing 200 ml 10% sodium hydroxide solution. The solution was extracted with ethyl ether. The combined organic layers were washed with 10% sodium carbonate solution followed by saturated brine, then dried with anhydrous sodium sulfate, and concentrated to give the crude product. Finally, column chromatography eluted with benzene afforded the spectroscopically pure benzyl chloride as a white solid in a yield of 50%-70%. Structures and purity were confirmed by GC/MS and NMR.

Preparation of Benzyl Ethers. To a solution of sodium hydride in dry ethyl ether, the alcohol (1.1-1.3 mol relative to benzyl chloride) in 100 mL dry THF was added dropwise with constant stirring under nitrogen over a period of 30 min. After no more hydrogen evolved, a solution of 1.0-8.0 g (4.6-50.0 mmol) benzyl chloride in 50 mL ethyl ether or ethylene glycol dimethyl ether was added dropwise over a period of 30 min. The reaction mixture was stirred for 24 hr under nitrogen. The reaction mixture was poured into 200 ml ice water containing 10% ammonium chloride, and then extracted with ether. The combined organic layers were washed with saturated brine, dried over anhydrous sodium sulfate, and concentrated to give the crude product. Finally, purification by flash column chromatography afforded spectroscopically pure benzyl ethers.

Photosensitized Deprotection of Benzyl Ethers. A solution of 5.0 g benzyl ether, 4.0 mg (0.1 mmol) methylene green (or an equivalent weight of rose bengal, erythrosin, methylene blue, or eosin B) in 100 mL acetonitrile:water (1:1) was degassed with nitrogen for 30 min and then irradiated with a 120W GE Miser or 150W Sylvania indoor spotlight for 6-12 hr. After irradiation, the solution was saturated with sodium chloride and extracted with chloroform several times. The combined organic layers were washed with sodium bisulfite solution to remove benzaldehyde. Finally, flash column chromatography gave spectroscopically pure product. Structures and purity were confirmed by GC/MS and NMR and compared with the authentic materials.

Acknowledgments

We are grateful to the Environmental Protection Agency, Office of Pollution Prevention and Toxics, Design for the Environment Program, and the University of Connecticut Environmental Research Institute (PPRDC) for financial support of this research. We further thank the Warner Jenkinson company, St. Louis, Missouri, for a generous gift of FD&C and D&C Certified dyes. Finally, we thank Dr. William F. Bailey for stimulating suggestions and conversations, and Dr. Paul T. Anastas for his many words of encouragement and insight.

Literature Cited

1. Page, P. C. B.; Niel, M. B. v.; Prodger, J. C. *Tetrahedron* **1989**, *45*, 7643-7677.
2. Greene, T. W.; Wuts, P. G. M. *Protective Groups in Organic Synthesis*; 2nd Ed; John Wiley & Sons: New York, NY, 1991; pp 209-210.
3. Eliel, E. L.; Morris-Natschke, S. *J. Am. Chem. Soc.* **1984**, *106*, 2937-2942.
4. Eliel, E. L.; Lynch, J. E. *Tetrahedron Lett.* **1981**, *22*, 2855-2858.
5. Greene, T. W.; Wuts, P. G. M. *Protective Groups in Organic Synthesis*; 2nd Ed.; John Wiley & Sons: New York, NY, 1991; pp 49-53.
6. Weinreb, S. M.; Epling, G. A.; Comi, R.; Reitano, M. *J. Org. Chem.* **1970**, *40*, 1356-1358.
7. Hatch, R. P.; Shringarpure, J.; Weinreb, S. M. *J. Org. Chem.* **1978**, *43*, 4172-4177.
8. Platen, M.; Steckhan, E. *Tetrahedron Lett.* **1980**, *21*, 511-514.
9. Cohen, S. G.; Parola, A.; Parsons, G. H., Jr. *Chem. Reviews* **1973**, *73*, 141-161.
10. Eliel, E. L.; Lynch, J. E.; Kume, F.; Frye, S. V. *Organic Syntheses* **1987**, *65*, 215-223.

RECEIVED August 4, 1994

Chapter 6

A Photochemical Alternative to the Friedel–Crafts Reaction

George A. Kraus, Masayuki Kirihara, and Yusheng Wu

Department of Chemistry, Iowa State University, Ames, IA 50011

The Friedel-Crafts reaction, a widely used reaction in both industrial and academic laboratories, produces several byproducts which must be handled as pollutants. An alternative which involves the photochemically-mediated reaction of an aldehyde with a quinone is described. This alternative chemistry can be applied to direct syntheses of the ring systems of the benzodiazepines and benzoxepins.

One of the research themes under active investigation by our group is the development of a set of generic reactions which will ultimately replace reactions that generate pollutants. That is, the objective of our research is to prevent pollution at the source. It is important to realize that pollution prevention is a long range undertaking. Redesigning the landscape of industrial organic chemistry will require rethinking a number of fundamental assumptions. At the outset, one of the issues that we addressed was what type of reagent would best fit the objectives of pollution prevention. An ideal reagent would be recyclable and would not harm the environment. We concluded that either electricity or light would be the best reagents. We elected to use visible light, in part because we already had considerable experience with organic photochemistry and in part because we felt that it could emerge as the more economical of the two choices. Another consideration is the concept of atom economy (*1*). According to the concept of atom economy, the optimal reaction would be one in which every atom in the starting materials would become part of the products. Our group has developed an alternative to the traditional Friedel-Crafts reaction which uses visible light and is entirely consistent with the goals of atom economy.

Shown below are three pharmaceutically-important drugs, each currently

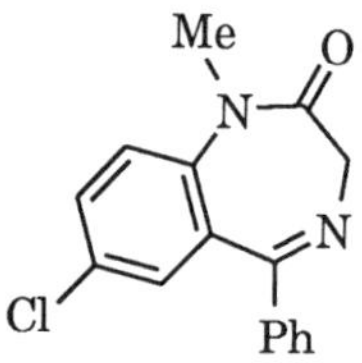

diazepam (Valium)

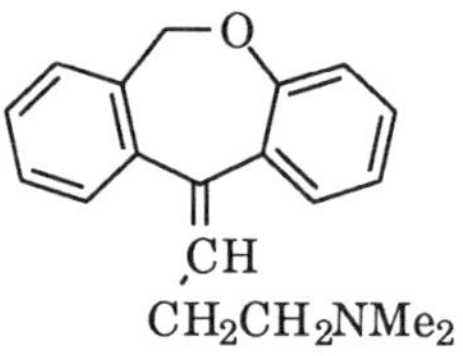

doxepin

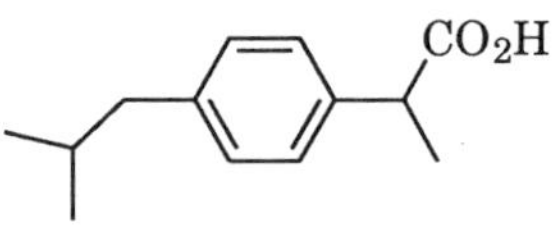

ibuprofen

0097–6156/94/0577–0076$08.00/0

produced annually in ton quantities (2). They are industrial objectives for which the Friedel-Crafts strategy has frequently been used. From the point of view of pollution prevention, there are a number of reasons why alternatives must be considered. The Friedel-Crafts pathway involves corrosive and air sensitive acid chlorides, Lewis acids such as aluminum chloride, stannic chloride or titanium tetrachloride and solvents such as nitrobenzene, carbon disulfide, carbon tetrachloride or methylene chloride (3). Although some research directed toward minimizing the amount of Lewis acid needed for the Friedel-Crafts reaction has been reported, this modification requires elevated temperatures (4). Friedel-Crafts reactions conducted with acidic resins have also been reported. This research appears promising but is in the early stages (5). Obviously, an attractive alternative to the traditional Friedel-Crafts reaction would have an impact on pollution prevention.

Experimental

Representative Procedure. Benzoquinone (4.80 g, 44.4 mmol) and freshly distilled butyraldehyde (20 mL, 346.7 mmol) were dissolved in dry benzene (240 mL) and were degassed with nitrogen for 15 min. The solution was irradiated with a high-pressure Hg-vapor lamp with a Pyrex filter for 5 days. The solution was concentrated in vacuo and the residue was purified by chromatography to afford 6.55g (82% yield) of **1a** as pale yellow crystals.

Spectral Data of Compounds from Table I. 2,5-Dihydroxyphenyl-1-butanone (1a): M.p. 94-96 °C (H-benzene). Lit.[7] m.p. 96 °C.
2,5-Dihydroxyphenyl (phenyl) methanone (1b): TLC (hexanes:ethyl acetate= 4:1) R_f = 0.38; mp 121-123 °C (H-benzene). Lit.[9] m.p.122-124 °C.
2,5-Dihydroxyphenyl-2-buten-1-one (1c): NMR (CDCl$_3$) d 1.97 (d, 3 H/2, J = 7 Hz), 2.02 (d, 3 H/2, J = 7 Hz), 6.01-6.07 (m, 1 H), 6.74-6.78 (m, 1 H), 6.87-7.05 (m, 2 H), 7.12-7.26 (m, 1 H), 12.27 (s, 1 H); IR (CH$_2$Cl$_2$) 1730, 1650, 1590 cm^{-1}; HRMS: m/z for C$_{10}$H$_{10}$O$_3$ calcd. 178.06299, measured 178.06273; TLC (hexanes:ethyl acetate= 4:1) R_f = 0.31; mp 114-116 °C (CHCl$_3$-EA).
3-Phenyl-1-(2,5-dihydroxyphenyl)-2-propen-1-one (1d) TLC (hexanes:ethyl acetate= 4:1) R_f = 0.32; mp 168-170 °C (H-benzene). Lit.[4] m.p. 170 °C.
2,5-dihydroxyphenyl (2-methoxyphenyl) methanone (1e): NMR (CDCl$_3$) d 3.80 (s, 3 H), 6.80 (s, 1 H), 6.92-7.09 (m, 4 H), 7.27-7.30 (m, 1 H), 7.49 (t, 1 H, J = 7 Hz), 11.75 (s, 1 H); IR (CH$_2$Cl$_2$) 1610 cm^{-1}; HRMS: m/z for C$_{14}$H$_{12}$O$_4$ calcd. 244.07356, measured 244.07312; TLC (hexanes:ethyl acetate= 4:1) R_f = 0.25; mp 145-148 °C (H-CHCl$_3$).
3-(2-Furanyl)-2,5-dihydroxyphenyl-2-propen-1-one (1f): NMR (CDCl$_3$) d 5.65 (s, 1 H), 6.50-6.51 (m, 2 H), 6.69 (d, 1 H, J = 3 Hz), 6.73-6.78 (m, 2 H), 6.94-6.98 (m, 2 H), 7.53-7.57 (m, 2 H); IR (CH$_2$Cl$_2$) 1720, 1630 cm^{-1}; HRMS: m/z for C$_{13}$H$_{10}$O$_4$ calcd. 230.05791, measured 230.05763; TLC (hexanes:ethyl acetate = 4:1) R_f = 0.48; mp 134-136 °C (CHCl$_3$-EA).
(1,4-Dihydroxy-2-naphthyl)-1-butanone (4a): TLC (hexanes:ethyl acetate= 4:1) R_f = 0.50; mp 141-143 °C (H-benzene). Lit.[8] m.p. 143 °C.
(1,4-Dihydroxy-2-naphthyl)phenylmethanone (4b): NMR (CDCl$_3$) d 7.41-7.53 (m, 3 H), 7.56-7.71 (m, 4 H), 8.11 (d, J= 4 Hz, 1 H), 8.50 (d, J= 4 Hz, 1 H), 13.54 (s, 1 H). IR (CHCl$_3$) 1600 cm^{-1}; MS: CI (NH$_3$) 282 (M$^+$ + NH$_4$); TLC (hexanes:ethyl acetate = 4:1) R_f = 0.50; mp 124-126 °C (H-CHCl$_3$).
(1,4-Dihydroxy-2-naphthyl)-2-buten-1-one (4c): NMR (CDCl$_3$) d 2.04 (d, 3 H, J = 7 Hz), 7.01 (d, 1 H, J = 6 Hz), 7.08 (s, 1 H), 7.20-7.23 (m, 1 H), 7.57 (t, 1 H, J = 5 Hz), 7.68 (t, 1 H, J = 5 Hz), 8.14 (d, 1 H, J = 5 Hz), 8.47 (d, 1 H, J = 5 Hz), 14.29 (s, 1 H); IR (CH$_2$Cl$_2$) 1640, 1590 cm^{-1}; HRMS: m/z for C$_{14}$H$_{12}$O$_3$ calcd. 228.07864, measured 228.07854; TLC (hexanes:ethyl acetate = 4:1) R_f = 0.33; mp 183-186 °C (CHCl$_3$-EA).

Results and Discussion

The synthesis of the substituted benzophenone shown below provides an instructive example of what we are trying to do. In principle, it could be synthesized by several possible pathways. However, if one asked any group of organic chemists how to make this compound, the Friedel-Crafts pathway would be suggested as the first choice by almost everyone. *This speaks to the conceptual redirection that needs to be made if pollution prevention is to be successful.* Our alternative involves the use of a quinone such as benzoquinone, an aldehyde and visible light (6). If the benzoquinone needed for the reaction was generated by the method of Frost (7), the overall approach would truly be environmentally benign.

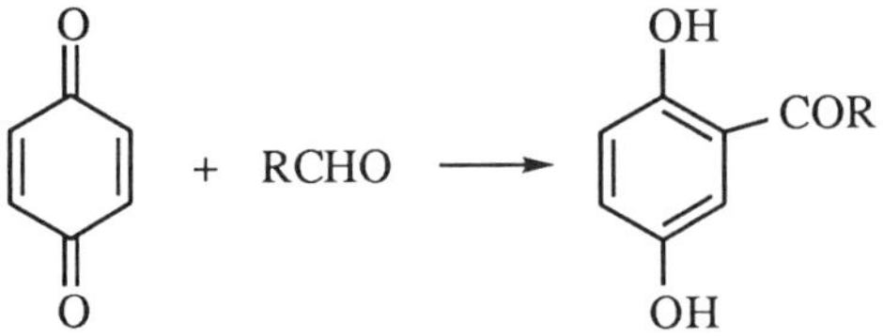

To the best of our knowledge, this reaction was discovered in the late 1800s by Klinger and Kolvenbach (8). Given the technology available at that time, the reaction was undoubtedly mediated by visible light. This photochemical reaction has been employed in recent years by Bruce (9) and by Maruyama (10). Last year we reported a study of the scope and limitations of this reaction from the point of view of the aldehyde component (6).

Table I shows reactions of aldehydes with benzoquinone and naphthoquinone. Some of the reactions have been conducted on a 20 gram scale. In more recent unpublished research, we examined the effects of ortho and para substitution. The

Table I. The Synthesis of 2-Acylhydroquinones

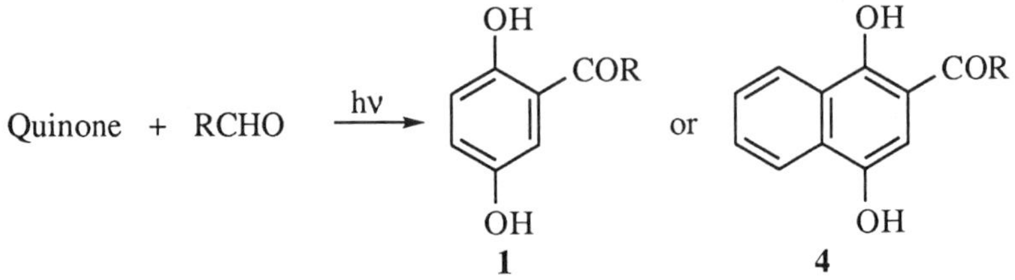

Entry	Quinone	R	% Yield	Product
1	2	Pr	82	1a
2	2	Ph	60	1b
3	2	$CH_3CH=CH$	52	1c
4	2	$PhCH=CH$	65	1d
5	2	$o\text{-}CH_3OC_6H_4$	62	1e
6	2	2-furyl-$CH=CH$	32	1f
7	3	Pr	77	4a
8	3	Ph	88	4b
9	3	$CH_3CH=CH$	65	4c

Quinone **2** is 1,4-benzoquinone and quinone **3** is 1,4-naphthoquinone.
SOURCE: Adapted from ref. 6.

yields shown are isolated yields after column chromatography. The successful results with the ortho isomer are significant, because the photochemistry of o-methyl

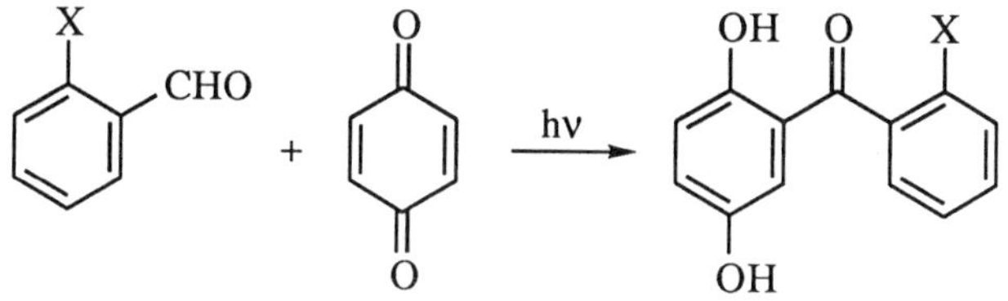

X =	Cl	78%
	OMe	72%
	Me	65%
	CO_2Me	58%

benzaldehyde is dramatically different from that of the para isomer. It is also significant from the point of view of synthetic planning, in that many industrially-significant drugs have substituents at the ortho-position on the aryl group. The compatibility of ester, ether and halide functionality is also noteworthy.

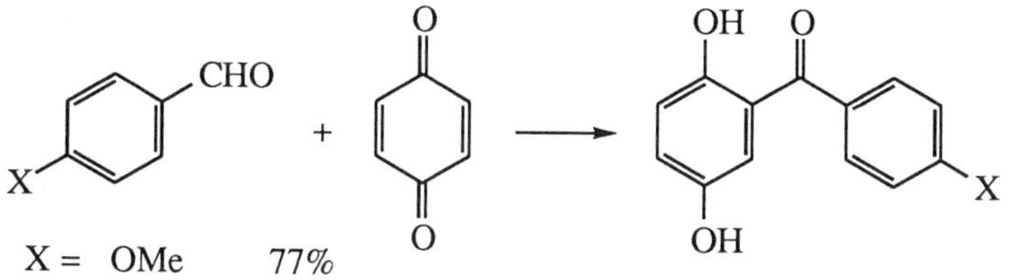

X =	OMe	77%
	Cl	73%
	CH_3	79%
	CHO	42%

As the photochemical reaction proceeds, a carbon-carbon bond is formed and a readily available aldehyde is transformed into an aromatic ketone. A possible mechanism is shown below. Electron transfer definitely plays a role here, but this mechanism adequately describes the reaction.

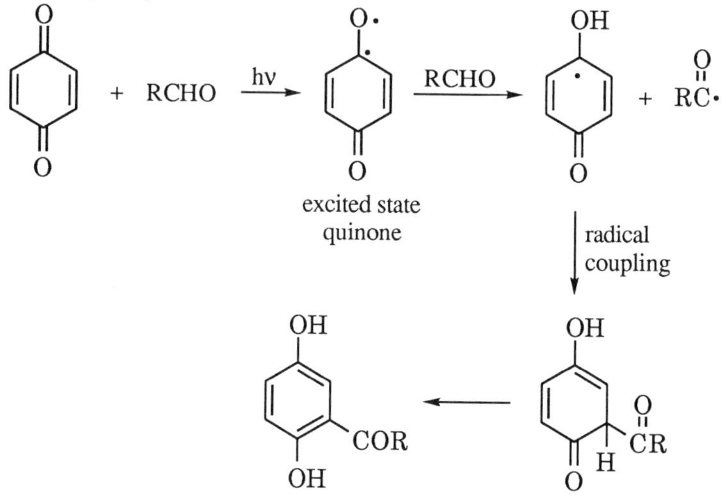

Applications to Diazepam Synthesis. The Sumitomo process for diazepam is depicted below (*11*). According to some reports, this is a major route by which diazepam (Valium), a widely-prescribed anti-anxiety drug, is prepared on an industrial scale. I have highlighted three reactions in the Sumitomo synthesis. These reactions involve toxic or corrosive reagents which would best be avoided if pollution control is an objective. It is worth noting that almost all of the other published pathways for the synthesis of diazepam involve Friedel Crafts reactions in the key carbon-carbon bond forming step.

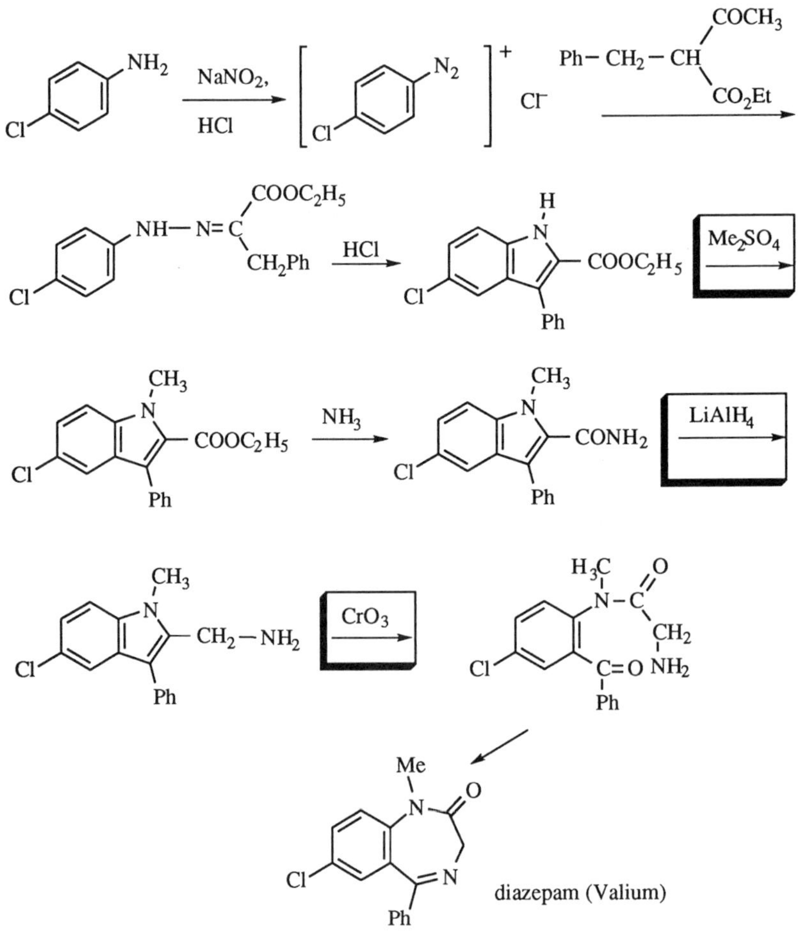

[Reaction scheme showing the Sumitomo synthesis of diazepam, starting from 4-chloroaniline, proceeding through diazotization with $NaNO_2$/HCl, coupling with $Ph-CH_2-CH(COCH_3)(CO_2Et)$, cyclization with HCl, N-methylation with Me_2SO_4, amidation with NH_3, reduction with $LiAlH_4$, oxidation with CrO_3, and final cyclization to diazepam (Valium).]

One of our alternative routes proceeds by way of the photochemically mediated addition of benzaldehyde to a benzoquinone bis-imine. Although the photochemically mediated reactions of aldehydes with imines have few precedents, we were pleased to find that the reaction of the bis-imine shown below with benzaldehyde afforded the desired product in 64% yield. The only drawback in this photochemical reaction was that the reduced benzoquinone bis-imine as produced. However, this appeared to be the only significant byproduct and of course this byproduct can be simply recycled to form more benzoquinone bis-imine. What remains to be accomplished in our synthetic route is the conversion of the amine into a chloride (generally accomplished using a diazotization reaction) and the appendage of the diazepine ring. The latter step has already been reported to proceed well.

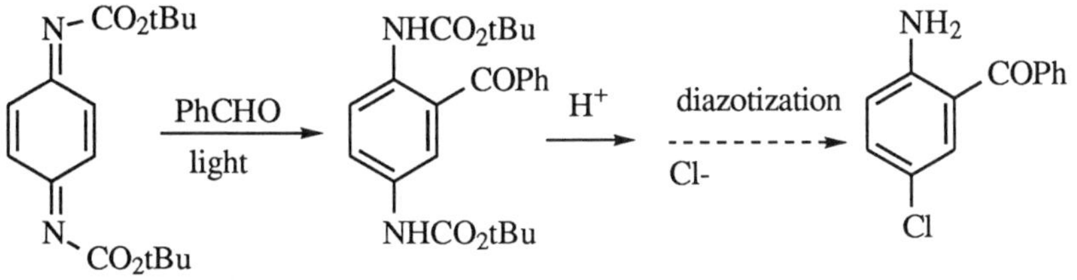

Perhaps a more efficient route to the benzodiazepine skeleton could be achieved if the compound shown below could be efficiently transformed into a benzodiazepine.

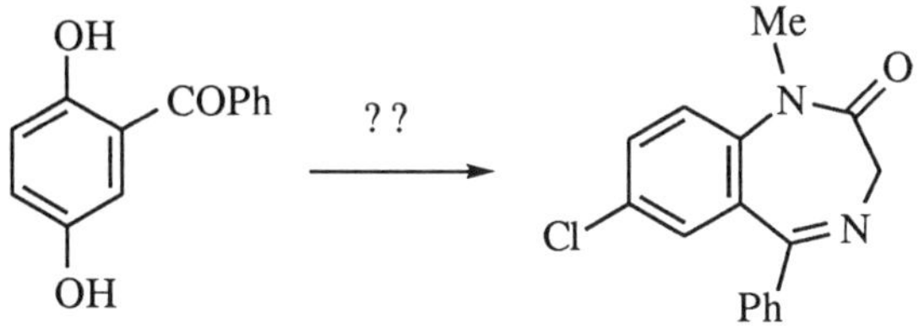

The benzophenone shown above has already been synthesized using our photochemical alternative to the Friedel-Crafts reaction. It was prepared by the reaction of benzoquinone and benzaldehyde and has been prepared on a 30 gram scale.

Transformation into the benzodiazepine skeleton requires the replacement of a carbon-oxygen bond by a carbon-nitrogen bond; that is, the conversion of a phenol into an aniline. Surprisingly, very few researchers have reported the successful conversion of a phenol into an aniline (*12*). We are presently not only examining the reported methods but also are working on the development of a new synthetic procedure.

An even more direct route would evolve from the connection of the two reagents shown below. Interestingly, the photochemically mediated reaction of a quinone with an imine has never been reported. If we can identify suitable reaction conditions for this transformation, we will have secured an exceptionally direct route to the benzodiazepine ring system. The imine required for this reaction has been previously synthesized and can readily be generated in multigram quantities by the condensation of benzaldehyde with the N-methyl amide of glycine. Even though the photochemical counterpart is not known, it is worth noting that hydrogen atom abstraction reactions of imines derived from benzaldehyde have been reported. Since hydrogen atom abstraction by an excited state of benzoquinone is one of the initial steps in the mechanism of the reaction, there is a very good possibility that this reaction will be successful.

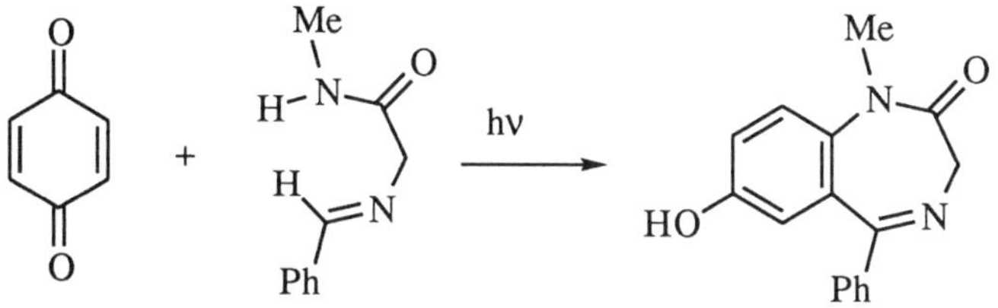

Applications to Doxepin Synthesis. The next synthetic objective to be discussed is doxepin (*13*). The structure of doxepin is shown below. Doxepin (also marketed as Adapin or Sinequan) is a tricyclic heterocyclic compound that is widely prescribed. Doxepin is prepared in very large quantities using a traditional Friedel-Crafts methodology. Of importance from the point of view of synthetic planning is that doxepin is sold as a mixture of E- and Z- double bond isomers. This leads to a simplification of the synthesis of doxepin.

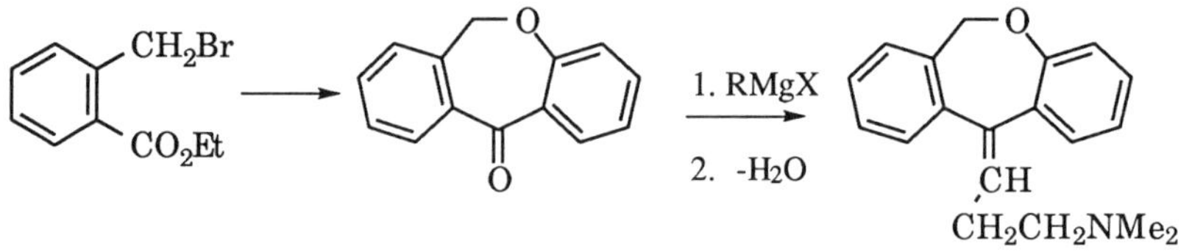

Our alternate synthetic route is shown below. Our route involves the photochemically mediated reaction between the methyl ester of ortho formyl benzoic acid and benzoquinone. We have already shown that this reaction is successful. Significantly, in one step we have generated the entire ring system present in doxepin.

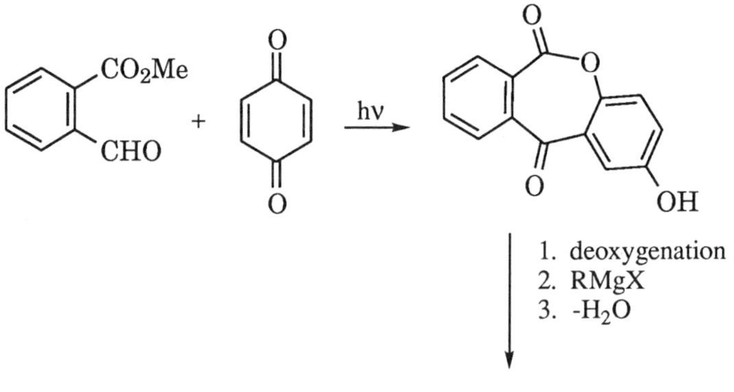

The transformation of the seven-membered ring lactone into an ether and the deoxygenation of the phenol both have extensive literature precedent. Once this is successful, we will then have prepared an advanced intermediate in the industrial synthesis of doxepin.

Conclusions

The production of toxic pollutants has become a problem for industry, academic research labs and even undergraduate labs. Unless innovative strategies are developed to prevent pollution at the source, the problem is certain to become more serious in the coming years. The idea of using visible light (a "reagent" which can be safely used in large excess) to develop a new generation of alternative synthetic pathways has many advantages. The theoretical framework for photochemical reactions is already well developed. Many general photochemical reactions are known and their modification for the use of visible light is often straightforward. Scale-up of photochemical reactions has precedent, since a few photochemical reactions are already used on large scale in industry. One of the most common involves chlorination or sulfonation by a radical process. A second photochemical reaction of major significance is the synthesis of vitamin D compounds from steroids. Photooxygenations involving singlet oxygen have also been used. For example, sensitized photooxygenation of citronellol is a key step in the industrial synthesis of rose oxide by Firmenich. Photopolymerization (for example by successive [2+2] cycloadditions) and photonitrosation are also industrially significant.

Our alternate synthesis strategy has the goal of preventing pollution at the source. In order for this goal to be implemented, realistic synthesis routes must be created. Pollution prevention operates at several levels. The most likely immediate benefit of any new synthetic reagents or synthetic reactions will be seen at the level of the research lab, where relatively small quantities of compounds are manipulated. Once the value of the new synthetic method is recognized and the scope and limitations of the reaction are clearly defined, the new method will gradually be used on larger and larger scales, eventually gaining acceptance at the pilot plant or industrial scale. Once the syntheses presented herein have been successfully scaled up, these syntheses will actually represent attractive alternatives. *If a commercial product such a diazepam or doxepin were prepared industrially using our methods, the environmental impact would be significant*, not only in terms of not having to transport and use tons of toxic reagents but also in terms of not having to dispose of large amounts of toxic byproducts. This research represents the initial phase of a concerted effort in pollution prevention. We intend to develop additional photochemical alternatives in the coming years.

Acknowledgments:
We are pleased to acknowledge the support of the U. S. Environmental Protection Agency, Office of Pollution Prevention and Toxics, Design for the Environment Program.

Literature Cited

1. Trost, B. M. *Science*, **1991**, *254*, 1471.
2. Pharmaceutical Chemistry, Drug Synthesis; Roth, H. J.; Kleemann, A.; T. Beisswenger, T., Eds. John Wiley: New York, New York, 1987; Vol.1.
3. Friedel-Crafts and Related Reactions; Olah, G. A., Ed. John. Wiley: New York, New York, 1964; pp 1-382.
4. Pearson, D. E.; Buehler, C. A. *Synthesis*, **1972**, 533.
5. NaFion catalyst: Olah, G. A., P.S.; Prakash, G.K.S. *Synthesis*, **1986**, 513.
6. Kraus, G. A.; Kirihara. M. *J. Org. Chem.* **1992**, *57*, 3256.
7. Dell, K. A.; Frost, J. W. *J. Am. Chem. Soc.*, **1993**, *115*, 11581.
8. Klinger, H.; Kolvenbach, W. *Chem. Ber.*, **1898**, *31*, 1214.
9. Bruce, J.M.; Creed, D.; Ellis, J.N. *J. Chem. Soc. (C)*, **1967**, 1486.
10. Maruyama, K.; Miyagi, Y. *Bull. Chem. Soc. Japan*, **1974**, *47*, 1303
11. Inaba, S.; Ishizumi, K.; Okamoto, T.; Yamamoto, H. *Chem. Pharm. Bull.* **1971**, *19*, 722.
12. Wynberg, H. *J. Org. Chem.*, **1993**, *58*, 5101.
13. Stach, K.; Bickelhaupt, F. *Monatsh.* **1962**, *93*, 896.

RECEIVED August 4, 1994

Chapter 7

Mn(III)-Mediated Electrochemical Oxidative Free-Radical Cyclizations

Barry B. Snider and Bridget A. McCarthy

Department of Chemistry, Brandeis University, Waltham, MA 02254–9110

Mn(III)-Based oxidative free-radical cyclizations and additions are versatile reactions that have been extensively developed over the past decade allowing free-radical cyclizations to be carried out without toxic tin reagents. Their applicability in large scale synthesis is limited by the requirement for ≥ 2 equivalents of $Mn(OAc)_3$ per mole of substrate oxidized. While manganese is not expensive, it is toxic, and the disposal of large quantities poses a pollution problem. A variety of procedures were investigated to develop a cheap and safe method for reoxidizing Mn(II) to Mn(III) *during* the oxidation of the organic substrate. Successful results in some cases were obtained using anodic oxidation to reoxidize Mn(II) to Mn(III) in a mixed solvent of AcOH and EtOH containing NaOAc as an electrolyte. Sodium periodate in DMSO is also capable of reoxidizing Mn(II) to Mn(III).

Free-radical cyclization of alkenes has become a valuable method for the synthesis of cyclic compounds during the past twenty years (1, 2). The most widely used procedure is the reduction of a halide or other functional group to a radical with R_3SnH, followed by cyclization and reduction of the resulting radical to a hydrocarbon in the chain propagation steps. This approach is seriously limited in several regards. First, an alkyl halide must be introduced into the starting material. Second, a relatively unfunctionalized product resulting from a net two-electron reduction is produced in the cyclization reaction (equation 1). Third, R_3SnH is used as a stoichiometric reagent producing very large quantities of highly toxic organic wastes.

Mn(OAc)₃-Based Oxidative Free-Radical Additions and Cyclizations

Oxidative free-radical cyclization, in which the initial radical is generated oxidatively, and/or the cyclic radical is oxidized to a cation or alkene to terminate the reaction has considerable synthetic and economic potential since more highly functionalized products can be prepared from simple precursors (equation 2). The oxidative addition of acetic acid to alkenes discovered by Heiba and Dessau (3, 4) and Bush and Finkbeiner (5) in 1968 provides the basis for a general procedure for carrying out oxidative free-radical cyclizations. Mn(OAc)₃ in acetic acid at reflux generates the carboxymethyl radical, which adds to alkenes to give a radical that is oxidized by a second equivalent of Mn(OAc)₃ to give a γ-lactone. This free-radical reaction efficiently forms a carbon-carbon bond and produces a highly functionalized product since the reaction is carried out under oxidative conditions. Although Heiba and Dessau continued to develop this area until 1974, organic chemists did not make much use of this reagent. However, in the past decade there has been an explosion of activity in development of oxidative free-radical reactions using Mn(OAc)₃ (6-25).

Oxidative Radical Generation — Cyclization — Oxidative Termination (2)

We have been interested in developing oxidative free-radical cyclizations using Mn(OAc)₃ (26-49). These reactions have the potential to prepare complex, highly functionalized, polycyclic molecules from simple precursors. They also pose a more stringent challenge for the method than oxidative additions of simple substrates, like acetic acid or acetone, to alkenes. Acetic acid and acetone are used in excess as solvent and the yield is based on the oxidant consumed. Further oxidation of the product is not generally a problem since the starting material is present in vast excess. In oxidative cyclizations, the substrate is too complex and expensive to use in excess. Further oxidation of the product is a major concern and common side reaction.

The synthesis of methyl *O*-podocarpate (**3**) is typical of the transformations that can be carried out with Mn(OAc)₃. Oxidation of **1** with 2 equivalents of Mn(OAc)₃ in AcOH at 15 °C afforded 50% of **2** as a single isomer that has been converted to methyl *O*-methylpodocarpate by Clemmensen reduction (26). Both enantiomers of *O*-methylpodocarpic acid have been prepared using the appropriate phenylmenthyl ester (49).

2, X = O
3, X = H₂

Tandem Oxidative Cyclization of 4 to Afford 8. The utility of tandem cyclizations in which both addends are alkenes was demonstrated in the synthesis of bicyclo[3.2.1]octanone **8** (*29, 36, 47*). Oxidation of **4** with 2 equivalents of Mn(OAc)$_3$ and 1 equivalent of Cu(OAc)$_2$ in acetic acid afforded 86% of **8**. Oxidation of **4** generated the α-keto radical **5** that underwent 6-*endo* cyclization to afford tertiary radical **6**. 5-*Exo* cyclization of 5-hexenyl radical **6** provided primary radical **7** as a 2:1 mixture of *exo* and *endo* isomers. Oxidation of both stereoisomers of **7** by Cu(OAc)$_2$ yielded alkene **8** and Cu(OAc), which was reoxidized to Cu(OAc)$_2$ by the second equivalent of Mn(OAc)$_3$. This process is therefore catalytic in Cu(OAc)$_2$ but consumes 2 equivalents of Mn(OAc)$_3$. Since this reaction proceeded selectively in very high yield, our initial goal was to make this reaction catalytic in Mn(OAc)$_3$ by either electrochemical or chemical oxidative regeneration of Mn(III) in the reaction mixture.

Indirect Electrochemical Synthesis

Electrochemical oxidations and reductions provide environmentally safe methods for carrying out organic synthesis. Anodic oxidation is the optimal technique for some oxidations, such as the Kolbe oxidation of carboxylic acids. However, many oxidations that can be carried out in high yield with the appropriate chemical oxidant cannot be accomplished by anodic oxidation. Indirect electrochemical oxidation provides a potential solution to this problem (*50, 51*). The reagent (mediator) carries out the oxidation of the substrate giving the product selectively and the reduced form of the mediator. The reduced form of the mediator is then oxidized electrochemically to generate the useful oxidized form of the mediator. The mediator is therefore used only in catalytic amounts. Indirect electrochemical oxidations and reductions thus have the potential to achieve the selectivity of chemical reactions with the environmental benefits of electrochemical methods.

Three types of indirect electrochemical reactions are known. External indirect electrochemical oxidation involves isolating the reacted form of the mediator and regenerating the useful form electrochemically in an external cell. In this method the chemical and electrochemical steps are run separately so that electrode processes do not interfere with the chemical reaction, nor do the substrate or product affect the electrode process. The disadvantage of this approach is that it is technically complicated and does not permit utilization of a continuous process.

The mediator and substrate are both placed in the cell in internal indirect electrochemical oxidation. This process permits the mediator to be used in catalytic amounts since it is recycled as shown for an oxidation in Figure 1. This approach facilitates purification since only a catalytic amount of the mediator is used and internal indirect electrochemical oxidation may permit the development of a continuous process. However, development of reaction conditions is very demanding since the substrate and product are subjected to electrode processes.

A third type of indirect electrochemical process involves processes in which the mediator is fixed at the electrode, such as 4-amino-2,2,6,6-tetramethylpiperidinyl-1-oxyl (TEMPO) modified graphite felt electrodes.

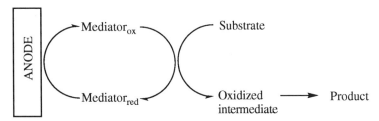

Figure 1. Indirect Electrochemical Oxidation

Recent Advances in Indirect Electrochemical Synthesis. Indirect electrochemical synthesis has tremendous potential as an environmentally benign procedure since the selectivity of a chemical reaction can be obtained without the production of toxic or hazardous byproducts. Two recent reviews (*50, 51*) exhaustively cover the field through the mid 1980's. Before describing our results, we will briefly survey recent advances in the field.

Reductions. Sm(II) has been developed as a versatile one-electron reductant of broad utility in organic synthesis. Sm(II) can be prepared and regenerated *in situ* by reduction of Sm(III) in DMF with a consumable magnesium anode (*52-54*). Under these conditions aromatic esters are reductively dimerized to 1,2-diketones with only 10% Sm(III) (*52*). Similarly, allylic chlorides can be added to ketones to give homoallylic alcohols (*53*). SmCl$_3$-catalyzed electrosynthesis of γ-butyrolactones from 3-chloro esters and ketones or aldehydes proceeds in 25-76% yield (*54*).

Electrocatalytic reductive coupling of aryl chlorides to afford biphenyls can be accomplished with dichloro(1,2-bis(di-propylphosphino)benzene)nickel(II) in yields as high as 96% with 2 mol % of the catalyst in polar, coordinating solvents (*55*). Similar couplings can also be achieved with nickel-2,2'-bipyridyl and Pd(PPh$_3$)$_2$Cl$_2$ as catalysts (*56, 57*). Indirect electrochemical reduction of vicinal dibromides to alkenes occurs efficiently with iron and cobalt porphyrins as mediators (*58*). Vitamin B$_{12}$ is a mediator for the indirect electrochemical reduction of α-halo acids (*59*).

Ketones can be reduced electrochemically in the presence of various mediators. Acetophenone is reduced to the alcohol at a lead cathode in the presence of SbCl$_3$ (*60*). Reduction of 6-hepten-2-one at a mercury cathode forms the alcohol. In the presence of dimethylpyrrolidinium tetrafluoroborate, the pinacol is formed in wet diglyme and the radical coupling product 1,2-dimethylcyclohexanol is formed in DMF (*61*).

A variety of reactions are catalyzed by electrochemically generated Ni(0) (*62*). Electrochemical reduction of Ni(bipy)$_3$Br$_2$ affords a reagent that couples acid chlorides and alkyl or aryl halides to form unsymmetrical ketones (*63*). Symmetrical ketones are formed from alkyl halides and carbon dioxide (*64*). Reductive electrochemical carboxylation of terminal alkynes, enynes and diynes can be accomplished with 10% Ni(bipy)$_3$(BF$_4$)$_2$ in DMF (*65-68*). Terminal alkynes lead selectively to α-substituted acrylic acids. Electrocatalytic hydrogenation on hydrogen-active electrodes has been reviewed (*69*). Radical cyclizations of vinyl, alkyl and aryl radicals can be carried out by indirect electrochemical reduction with a Ni(II) complex as a mediator (*70*).

Oxidations. Electrocatalytic oxidative coupling of 2-naphthols to 1,1'-binaphthyls in quantitative yield in more than 90% current efficiency can be accomplished on a graphite felt electrode coated with a thin poly(acrylic acid) layer immobilizing TEMPO (*71*). Similar oxidative couplings have been carried out on methylquinolines (*72*). Indirect electrochemical oxidations of alcohols to carbonyl compounds have been achieved with TEMPO and electrochemically generated tribromide or other active bromide species (*73, 74*), polypyridine Ru(IV) species (*75*) and polybromochloride ions (*76*). Halonium ions generated electrochemically have been used for the halogenation of pyrimidines (*77, 78*), haloform reactions of methyl ketones (*79*) and α-methoxylation of protected amino acids (*80*).

Indirect electrochemical oxidative carbonylation with a palladium catalyst converts alkynes, carbon monoxide and methanol to substituted dimethyl maleate esters (*81*). Indirect electrochemical oxidation of dienes can be accomplished with the palladium-hydroquinone system (*82*). Olefins, ketones and alkylaromatics have been oxidized electrochemically using a Ru(IV) oxidant (*83, 84*). Indirect electrooxidation of alkylbenzenes can be carried out with cobalt, iron, cerium or manganese ions as the mediator (*85*). Metalloporphyrins and metal salen complexes have been used as mediators for the oxidation of alkanes and alkenes by oxygen (*86-90*). Reduction of oxygen and the metalloporphyrin generates an oxoporphyrin that converts an alkene into an epoxide.

Mn(III)-Mediated Indirect Electrochemical Oxidations

At the current level of development, Mn(III)-based oxidative free-radical cyclization is a very attractive procedure. Simple substrates are converted readily to highly functionalized and versatile products. The preferred solvent, acetic acid, is relatively safe since it becomes vinegar on dilution with water. Cu(OAc)$_2$ is used in catalytic quantities. However, 2 equivalents of Mn(OAc)$_3$ must be used. The amount of Mn(II) waste would be significantly decreased if Mn(OAc)$_3$ could be used in catalytic quantities and regenerated by *in situ* reoxidation of the Mn(II) produced in the reaction. From the pollution point of view, electrochemical oxidation would be an effective way to regenerate Mn(III). Oxidative free-radical cyclization using Mn(III) and Cu(II) as catalysts in an electrochemical oxidation would minimize the production of toxic chemicals and the resulting pollution.

Side chain oxidation of aromatics by Mn(III)-mediated indirect electrochemical oxidation is well known (*50, 51, 85*). Several groups have recently examined Mn(OAc)$_3$-mediated electrochemical oxidations. Coleman *et. al.*, at Monsanto have developed a procedure for the synthesis of sorbic acid precursors from acetic acid and butadiene using electrochemical mediated Mn(OAc)$_3$ oxidation at 100 °C in acetic acid under pressure with graphite plates or a graphite felt anode (*91*). These authors concluded that this approach was practical for the large scale synthesis of sorbic acid.

Nishiguchi has shown that electrochemical mediated Mn(OAc)$_3$ oxidation can be used to add acetic acid to styrenes at 95-97 °C under constant current conditions in a beaker-type divided cell with carbon rods as anode and cathode and a ceramic cylinder as a diaphragm (92-95). The addition of ethyl cyanoacetate to alkenes occurs in the same type of cell at 40 °C. These reactions were carried out with 0.2 equivalents of Mn, which is 10% of what would be needed in a stoichiometric reaction. Nédélec used electrochemically generated Mn(III) to initiate the radical chain addition of dimethyl bromomalonate to alkenes (96).

Mn(III)-Mediated Electrochemical Oxidative Free-Radical Cyclizations. The results of Coleman and Nishiguchi suggest that it may be possible to develop Mn(III)-mediated electrochemical oxidative cyclizations. However, we have previously found that oxidation of complex substrates, which cannot be used in excess because of their expense, poses a problem that does not have to be dealt with when the solvent, acetic acid, is being oxidized. In the work of Coleman and Nishiguchi discussed above yields are based on the current passed through the solution, not on the amount of starting materials used. In the cyclization reactions of complex and expensive substrates, yields must be based on the amounts of starting materials used and not on the current passed through the solution. This poses a more rigorous test for the electrochemical reoxidation method.

Electrochemical Oxidative Cyclization of 4. We chose to use **4** for initial studies because it reacted with 2 equivalents of Mn(OAc)$_3$ and 1 equivalent of Cu(OAc)$_2$ to give **8** in 86% yield (36). The stoichiometric oxidative cyclization requires several hours at room temperature, but proceeds much faster at higher temperatures. The electrochemical reactions must be carried out at elevated temperatures since the concentration of Mn(III) will be much lower. Careful optimization of temperature and current will be required so that Mn(III) is consumed by oxidation of **4** at the same rate that it is regenerated by electrochemical oxidation of Mn(II). We chose to carry out initial trials at 60 °C as shown in Table I.

Table I. Mn(III)-Mediated Electrochemical Oxidative Cyclization of 4

Run	Eq. Mn	Eq. Cu	Solvent	Temp (°C)	Current (mA)	Charge (coul.)[a]	8 (%)	unreacted 4 (%)
1	0.2	0.1	AcOH	60	80	882	13	7
2	0.2	0.1	EtOH	60	80	570	2	90
3	0.2	0.1	4:1 AcOH/EtOAc	60	80	520	20	8
4	0.2	0.1	4:1 AcOH/EtOH	60	80	456	43	10
5	0.2	0.1	3:2 AcOH/EtOH	60	80	522	15	5
6	0.2	0.1	9:1 AcOH/EtOH	60	80	452	18	9
7	0.2	0.1	4:1 AcOH/EtOH	60	50	390	31	6
8	0.2	0.1	4:1 AcOH/EtOH	60	20	700	15	5
9	0.2	0.1	4:1 AcOH/EtOH	80	80	908	25	10
10	0.2	0.1	4:1 AcOH/EtOH	40	80	1490	20	6
11	0.4	0.1	4:1 AcOH/EtOH	60	80	524	56	9
12	0.1	0.1	4:1 AcOH/EtOH	60	80	520	8	70
13	0.2	0.1	4:1 AcOH/EtOH[b]	60	80	1560	59	12
14	0.2	0.1	4:1 AcOH/EtOH[c]	60	80	1175	56	9
15	0.0	0.1	4:1 AcOH/EtOH	60	80	1000	0	82

[a]Theoretical charge required is 386 coulombs. [b]Contains 2% water. [c]Contains 4% water

Electrochemical oxidations were carried out in a divided H-type cell with a fine porosity glass frit as the diaphragm and 1/4" graphite rods as electrodes. All oxidations were carried out with 2 mmol of **4** in 0.05 M solution with 0.75 M NaOAc as the electrolyte at constant current in the solvents indicated in Table I. Mn(OAc)$_2$ and Cu(OAc)$_2$ were used in the amounts indicated. Since 2 equivalents of Mn(OAc)$_3$ are required, 0.2 equivalents is 10% of the theoretical amount. No **8** is formed in the absence of current since Mn(II) was used; all the Mn(III) generated is formed electrochemically. Since 2.0 F/mol of electricity is required for two one-electron oxidations of **4**, the reaction theoretically should be complete when 386 coulombs have been passed through the solution.

Although AcOH is the most common solvent for Mn(OAc)$_3$ oxidations, these reactions can also be carried out in EtOH, MeOH, DMF, DMSO and CH$_3$CN (*37, 47*). Initial results (runs 1 and 2) indicated that low yields of **8** were obtained in either AcOH or EtOH. Nishiguchi has used AcOH/EtOAc mixtures for inter-molecular Mn(III)-mediated electrochemical reactions (*92-95*). This solvent mixture was not efficacious for the oxidative cyclization of **4** (run 3). We were delighted to find that modest yields of **8** were obtained in 4:1 AcOH/EtOH with good current efficiency (run 4). Lower yields were obtained in 3:2 or 9:1 AcOH/EtOH (runs 5 and 6), at lower current densities (runs 7 and 8) and at higher or lower temperatures (runs 9 and 10). There was a modest increase in yield with 0.4 equivalents of Mn(II) (run 11) and a significant decrease in yield with 0.1 equivalents of Mn(II) (run 12).

The yield of **8** was improved to 56-59% by adding 2-4% water to the solution (runs 13 and 14). While the yield of **8** increased significantly in the presence of water, the current efficiency dramatically decreased since ≈10% unreacted **4** was still present after 6-8 F/mol of electricity had passed through the solution. Addition of 2-4% water to the solution decreased the applied voltage at 80 mA from 35 V to 19 V.

These results indicate that Mn(III)-mediated electrochemical oxidative cyclization of **4** to give **8** proceeded in 43-59% yield with only 10% of the theoretical amount of Mn(II). Control experiments established that this is not a direct electrochemical oxidation because no **8** was obtained in the absence of Mn(II) (run 15). However, the yields were not comparable to the 86% yield of **8** obtained with 2 equivalents of Mn(III). We are not sure why the yield of **8** was lower in the electrochemical oxidation, but we have established that **8** is stable to the reaction conditions and does not react further.

Mn(III)-Mediated Electrochemical Oxidative Cyclization of 9. We next turned our attention to other substrates using the optimized conditions from runs 4 and 13 of Table I. Tandem oxidative cyclization of β-Keto ester **9** with 2 equivalents of Mn(OAc)$_3$ and 1 equivalent of Cu(OAc)$_2$ in acetic acid afforded 72% of **10** and 3% of a *cis*-fused diastereomer (*36, 40*). Mn(III)-mediated electrochemical oxidation of **9** in 4:1 AcOH/EtOH at 60 °C provided only 15% of **10** and 15% of the monocyclic compound **12**, which was formed by Cu(II) oxidation of the monocyclic radical **11**. A similar reaction in a solvent mixture containing 2% water afforded 7% of **10** and 3% of **12**. The formation of **12** under these conditions was particularly surprising since the oxidation of **11** should be favored at high, not low, Cu(II) concentrations (*39*). No **12** was formed in the stoichiometric oxidation with 1 equivalent of Cu(II), yet equal amounts of **10** and **12** are formed in the electrochemical oxidation with 0.1 equivalents of Cu(II).

Mn(III)-Mediated Electrochemical Oxidative Cyclization of 13. Oxidative cyclization of β-Keto ester **13** with 2 equivalents of Mn(OAc)$_3$ and 1 equivalent of Cu(OAc)$_2$ in acetic acid at room temperature yielded 56% of **14**, 3% of the diastereomer **15** and 14% of the trisubstituted alkene **16** (*35*). Mn(III)-

mediated electrochemical oxidation of **13** in 4:1 AcOH/EtOH at 60 °C provided only 19% of **14**, 4% of **16**, 8% of **18** and 1% of **19**. A similar reaction in a solvent mixture containing 2% water afforded 17% of **14**, 3% of **16** and no **18** or **19**.

The unexpected products **18** and **19** were formed by further oxidation of the major product **14**, which occurred at 60 °C, but not at room temperature. Tandem oxidative cyclization of β-Keto ester **13** with 4 equivalents of Mn(OAc)$_3$ and 1 equivalent of Cu(OAc)$_2$ in acetic acid at room temperature provided only **14-16**. However **18** was the major product from oxidation of **13** with 4 equivalents of Mn(OAc)$_3$ and 1 equivalent of Cu(OAc)$_2$ at 60 °C. The intermediacy of **14** was established by oxidation of **14** with 2 equivalents of Mn(OAc)$_3$ and 1 equivalent of Cu(OAc)$_2$ in AcOH at 60 °C to provide 70% of a 10:1 mixture of **18** and **19**. Oxidation of **14** will afford α-keto radical **17**, which will be oxidized to give **19** or cyclize to give a bicyclic radical that will be oxidized by Cu(II) to give **18**. The formation of **18** and **19** in the Mn(III)-mediated electrochemical oxidation of **13** thus results solely from the necessity of using a higher reaction temperature to obtain a reasonable rate of reaction at the lower Mn(III) concentration of Mn(III)-mediated electrochemical oxidative cyclization.

Mn(III)-Mediated Electrochemical Oxidation of 20, 22 and 24. Oxidative cyclization of **20** with 4 equivalents of Mn(OAc)$_3$ and 1 equivalent of Cu(OAc)$_2$ afforded 78% of methyl salicylate (**21**) (*32*). Mn(III)-mediated electrochemical oxidation of **20** in 4:1 AcOH/EtOH at 60 °C provided only recovered starting material after passage of 4 F/mol of electricity. Similarly, **22**, which gave

71% of **23** with 2 equivalents of Mn(OAc)$_3$ and 1 equivalent of Cu(OAc)$_2$ (*35*), provided no **23** after passage of 6 F/mol of electricity through the system. Mn(III)-mediated electrochemical cyclization of **24** yielded 11% of **25** in contrast to the 48% yield obtained with 2 equivalents of Mn(OAc)$_3$ and 1 equivalent of Cu(OAc)$_2$ (*31*).

It is not clear why the Mn(III)-mediated electrochemical cyclization of **20** and **22** was completely unsuccessful. It should be noted that the rate determining step in the cyclization of α-unsubstituted β-keto esters is the addition of the alkene to the Mn(III) enolate with loss of Mn(II) while the rate determining step in the oxidation of α-alkyl substituted β-keto esters is the formation of the Mn(III) enolate, which rapidly loses Mn(II) to form the free α-keto radical (*31, 40*). Mn(III)-mediated electrochemical oxidation was somewhat successful for α-alkyl substituted β-keto esters, but failed for α-unsubstituted β-keto esters.

Chemical Reoxidation of Mn(II) to Mn(III)

A variety of chemical oxidants might conceivably reoxidize Mn(II) to Mn(III). Potassium persulfate is inexpensive and easily handled since there is a large kinetic barrier to homolytic dissociation (*97, 98*). It is a very powerful oxidant that produces reactive SO$_4$·$^-$, which does not oxidize most organic functional groups but will oxidize metals such as Ag(I). Bromate, periodate and hydrogen peroxide have been used to oxidize Mn(II) in aqueous solution in studies of oscillating reactions (*99-102*). Hydrogen peroxide has been used to oxidize Mn(II) in biochemical model studies (*103*). *t*-Butyl hydroperoxide has been used to generate a Mn catalyst for oxidation of alkanes (*104*). Peracetic acid has recently been reported to reoxidize Mn(II) to Mn(III) in acetic acid (*105*).

As in the Mn(III)-mediated electrochemical oxidations, we chose to use **4** for initial studies because the reaction with 2 equivalents of Mn(OAc)$_3$ and Cu(OAc)$_2$ to give **8** proceeded cleanly in high yield. Oxidation of **4** in DMSO with 0.1 equivalents (5% of theoretical) of Mn(OAc)$_3$, 0.2 equivalents of Cu(OAc)$_2$ and 2 equivalents of NaIO$_4$ in DMSO at 60 °C for 2 d gave 60% of **8**. Slightly lower yields of **8** were obtained in AcOH. Complex mixtures of **8** and other products were obtained in DMF and EtOH. Oxidation of **4** with 0.2 equivalents (10% of theoretical) of Mn(OAc)$_3$, 0.2 equivalents of Cu(OAc)$_2$ and 2 equivalents of K$_2$S$_2$O$_8$ in DMSO for 1 d at 80 °C followed by addition of 2 additional equivalents of K$_2$S$_2$O$_8$ and heating for 1 d at 100 °C yielded 38% of **8**. Complex mixtures were obtained in AcOH and MeOH. Use of KBrO$_3$ to reoxidize Mn(II) resulted in destruction of **4** and production of complex mixtures. Although peracetic acid has recently been reported to reoxidize Mn(II) to Mn(III) in acetic acid (*105*), it epoxidized the disubstituted double bond of **4** more rapidly than it oxidized Mn(II).

Conclusion.

Mn(III)-mediated electrochemical oxidative cyclization was successful in some cases, such as that of **4**, which gives up to 59% of **8** with only 10% of the theoretical amount of Mn(III). However, the electrochemical cyclizations generally give lower yields than the stoichiometric Mn(III) oxidative cyclizations, and more complex product mixtures are typically obtained. One problem is that Mn(III) oxidations are relatively slow at room temperature. Therefore elevated temperatures must be used in the electrochemical cyclizations. This can lead to further oxidation as in the formation of **18** and **19** from **9** by oxidation of the initial product **14**. It is hard to determine why the yields are lower in other cases since we were unable to isolate byproducts. We can only speculate that either the starting materials or the products are oxidized at the electrode to give polymeric material. NaIO$_4$ in DMSO can be used to chemically reoxidize Mn(II) to Mn(III). Under these conditions **4** affords 60% of **8** with only 5% of the theoretical amount of Mn(III).

Mn(III)-based oxidative free-radical cyclization is an attractive alternative to tin hydride based procedures both in terms of chemical efficiency and pollution prevention potential since toxic tin byproducts are not produced and Mn(II) can be reoxidized to Mn(III) by $KMnO_4$. Mn(III)-mediated electrochemical oxidative cyclization has significant potential for pollution prevention since only catalytic amounts of Mn(III) are needed. Our results indicate that each case must be examined in detail. Some substrates are oxidized in high yield under the electrochemical conditions; others give very low yields of products. Further work is needed to understand these differences and develop more general catalytic conditions.

Indirect electrochemical synthesis is an area of much current interest as indicated in the introduction to this chapter. This field has tremendous potential for carrying out reactions in high yield under reaction conditions that minimize the formation of polluting byproducts.

Experimental Section

Oxidative Cyclization of 4 (Run 13). To a solution of β-keto ester **4** (420 mg, 2 mmol) in 30 mL of 0.75 M NaOAc in 78:20:2 AcOH:EtOH:H_2O was added a solution of $Mn(OAc)_2$ (69 mg, 0.4 mmol) and $Cu(OAc)_2 \cdot H_2O$ (40 mg, 0.2 mmol) in 10 mL of 78:20:2 AcOH:EtOH:H_2O. The mixture was placed into the anodic chamber of a glass H-type cell equipped with 1/4" graphite rod electrodes with a fine porosity glass frit as the diaphragm and stirred magnetically under N_2 at 60 °C. The cathodic chamber was filled with 15 mL of 0.75 M NaOAc in the same solvent mixture. Electrochemical oxidation was carried out with an Electrosynthesis Company model 420A power supply with an output of 500 mA at 70V attached to a model 410 potentiostat/galvanostat controller. A model 640 digital coulometer was used to measure the amount of current which passed through the system. After 1560 coulombs had passed through the system, TLC indicated that **4** had reacted completely. The anolyte was poured into 50 mL of water and extracted with ether (3×30 mL). The combined ether extracts were washed with saturated $NaHCO_3$ and NaCl solutions and dried ($MgSO_4$). Removal of the solvent under reduced pressure gave 323 mg (77%) of a yellow oil containing **4** and **8**. Flash chromatography (9:1 hexane-EtOAc) afforded 49 mg of recovered **4** (12%) followed by 248 mg of **8** (59%). The aqueous layer remaining after ether extraction was extracted with CH_2Cl_2 (3×30 mL). The combined CH_2Cl_2 layers were washed with saturated $NaHCO_3$ and brine solutions, dried ($MgSO_4$) and concentrated to give 48 mg (11%) of uncharacterizable oligomer.

Extraction of reaction mixtures with CH_2Cl_2 gave >90% material balance containing both monomeric and oligomeric material. The polar oligomeric material was easily removed by flash chromatography. Extraction with ether gave a lower material balance containing 80-90% pure monomeric material.

Acknowledgments. We are grateful to the U.S. Environmental Protection Agency, Office of Pollution Prevention and Toxics, Design for the Environment Program for financial support.

Literature Cited

1. Curran, D. P. *Synthesis* **1988**, 417 and 489.
2. Geise, B. *Radicals in Organic Synthesis: Formation of Carbon-Carbon Bonds;* Pergamon: Oxford, 1986.
3. Heiba, E. I.; Dessau, R. M.; Koehl, W. J. *J. Am. Chem. Soc.* **1968**, *90*, 5905.
4. Heiba, E. I.; Dessau, R. M.; Williams, A. L.; Rodewald, P. G. *Org. Synth.* **1983**, *61*, 22.

5. Bush, J. B.; Finkbeiner, H. *J. Am. Chem. Soc.* **1968**, *90*, 5903.
6. de Klein, W. J. In *Organic Synthesis by Oxidation with Metal Compounds*; Mijs, W. J.; de Jonge, C. R. H., Eds.; Plenum Press: New York, 1986; pp 261-314.
7. Badanyan, Sh. O.; Melikyan, G. G.; Mkrtchyan, D. A. *Russ. Chem. Rev.* **1989**, *58*, 286; *Uspekhi Khimii* **1989**, *58*, 475.
8. Melikyan, G. G. *Synthesis*, **1993**, 833.
9. Fristad, W. E.; Peterson, J. R. *J. Org. Chem.* **1985**, *50*, 10.
10. Fristad, W. E.; Hershberger, S. S *J. Org. Chem.* **1985**, *50*, 1026.
11. Fristad, W. E.; Peterson, J. R.; Ernst, A. B. *J. Org. Chem.* **1985**, *50*, 3143.
12. Fristad, W. E.; Peterson, J. R.; Ernst, A. B.; Urbi, G. B. *Tetrahedron* **1986**, *42*, 3429.
13. Yang, F. Z.; Trost, M. K.; Fristad, W. E. *Tetrahedron Lett.* **1987**, *28*, 1493.
14. Peterson, J. R.; Egler, R. S.; Horsley, D. B.; Winter, T. J. *Tetrahedron Lett.* **1987**, *28*, 6109.
15. Corey, E. J.; Kang, M. *J. Am. Chem. Soc.* **1984**, *106*, 5384.
16. Corey, E. J.; Gross, A. W. *Tetrahedron Lett.* **1985**, *26*, 4291.
17. Corey, E. J.; Ghosh, A. K. *Tetrahedron Lett.* **1987**, *28*, 175.
18. Corey, E. J.; Ghosh, A. K. *Chem. Lett.* **1987**, 223.
19. Surzur, J. M.; Bertrand, M. P. *Pure Appl. Chem.* **1988**, *60*, 1659.
20. Oumar-Mahamat, H.; Moustrou, C.; Surzur, J. M.; Bertrand, M. P. *Tetrahedron Lett.* **1989**, *30*, 331.
21. Oumar-Mahamat, H.; Moustrou, C.; Surzur, J. M.; Bertrand, M. P. *J. Org. Chem.* **1989**, *54*, 5684.
22. Bertrand, M. P.; Surzur, J. M.; Oumar-Mahamat, H.; Moustrou, C. *J. Org. Chem.* **1991**, *56*, 3089.
23. Citterio, A.; Cerati, A.; Sebastiano, R.; Finzi, C. *Tetrahedron Lett.* **1989**, *30*, 1289.
24. Citterio, A.; Fancelli, D.; Finzi, C.; Pesce, L.; Santi, R. *J. Org. Chem.* **1989**, *54*, 2713.
25. Citterio, A.; Santi, R.; Fiorani, T.; Strologo, S. *J. Org. Chem.* **1989**, *54*, 2703.
26. Snider, B. B.; Mohan, R. M.; Kates, S. A. *J. Org. Chem.* **1985**, *50*, 3659.
27. Snider, B. B.; Mohan, R. M.; Kates, S. A. *Tetrahedron Lett.* **1987**, *28*, 841.
28. Mohan, R.; Kates, S. A.; Dombroski, M.; Snider, B. B. *Tetrahedron Lett.* **1987**, *28*, 845.
29. Snider, B. B.; Dombroski, M. A. *J. Org. Chem.* **1987**, *52*, 5487.
30. Merritt, J. E.; Sasson, M.; Kates, S. A.; Snider, B. B. *Tetrahedron Lett.* **1988**, *29*, 5209.
31. Snider, B. B.; Patricia, J. J.; Kates, S. A. *J. Org. Chem.* **1988**, *53*, 2137.
32. Snider, B. B.; Patricia, J. J. *J. Org. Chem.* **1989**, *54*, 38.
33. Snider, B. B.; Buckman, B. O. *Tetrahedron* **1989**, *45*, 6969.
34. Snider, B. B.; Kwon, T. *J. Org. Chem.* **1990**, *55*, 1965.
35. Kates, S. A.; Dombroski, M. A.; Snider, B. B. *J. Org. Chem.* **1990**, *55*, 2427.
36. Dombroski, M. A.; Kates, S. A.; Snider, B. B. *J. Am. Chem. Soc.* **1990**, *112*, 2759.
37. Snider, B. B.; Merritt, J. E.; Dombroski, M. A.; Buckman, B. O. *J. Org. Chem.* **1991**, *56*, 5544.
38. Snider, B. B.; Wan, B. Y.-F.; Buckman, B. O.; Foxman, B. M. *J. Org. Chem.* **1991**, *56*, 328.
39. Snider, B. B.; Merritt, J. E. *Tetrahedron* **1991**, *47*, 8663.
40. Curran, D. P.; Morgan, T. M.; Schwartz, C. E.; Snider, B. B.; Dombroski, M. A. *J. Am. Chem. Soc.* **1991**, *113*, 6607.

41. Snider, B. B.; Buckman, B. O. *J. Org. Chem.* **1992**, *57*, 322.
42. Dombroski, M. A.; Snider, B. B. *Tetrahedron* **1992**, *48*, 1417.
43. Snider, B. B.; Zhang, Q.; Dombroski, M. A. *J. Org. Chem.* **1992**, *57*, 4195.
44. Snider, B. B.; Zhang, Q. *Tetrahedron Lett.* **1992**, *33*, 5921.
45. Snider, B. B.; Zhang, Q. *J. Org. Chem.* **1993**, *58*, 3185.
46. Snider, B. B.; McCarthy, B. A. *Tetrahedron* **1993**, *49*, 9447.
47. Snider, B. B.; McCarthy, B. A. *J. Org. Chem.* **1993**, *58*, 6217.
48. Snider, B. B.; Vo, N. H.; Foxman, B. M. *J. Org. Chem.* **1993**, *58*, 7228.
49. Zhang, Q.; Mohan, R.; Cook, L.; Kazanis, S.; Peisach, D.; Foxman, B. M.; Snider, B. B. *J. Org. Chem.* **1993**, *58*, 7640.
50. Torii, S. *Electro-organic Syntheses, Methods and Applications, Part I: Oxidations*; VCH: Weinheim, Germany, 1985; Chapter 11.
51. Steckhan, E. *Angew. Chem., Int. Ed. Engl.* **1986**, *25*, 683.
52. Hébri, H.; Duñach, E.; Heintz, M.; Troupel, M.; Périchon, J. *Synlett.* **1991**, 901.
53. Hebri, H.; Duñach, E.; Périchon, J. *Tetrahedron Lett.* **1993**, *34*, 1475.
54. Hebri, H.; Duñach, E.; Périchon, J. *J. Chem Soc., Chem. Commun.* **1993**, 499.
55. Fox, M. A.; Chandler, D. A.; Lee, C. *J. Org. Chem.* **1991**, *56*, 3246.
56. Meyer, G.; Rollin, Y.; Perichon, J. *J. Organometal. Chem.* **1987**, *333*, 263.
57. Jutand, A.; Négri, S.; Mosleh, A. *J. Chem. Soc., Chem. Commun.* **1992**, 1729.
58. Lexa, D.; Savéant, J.-M.; Schäfer, H. J.; Su, K.-B.; Vering, B.; Wang, D. L. *J. Am. Chem. Soc.* **1990**, *112*, 6162.
59. Rusling, J. F.; Miaw, C. L.; Couture, E. C. *Inorg. Chem.* **1990**, *29*, 2025.
60. Ikeda, Y.; Manda, E. *Chem. Lett.* **1989**, 839.
61. Swartz, J. E.; Mahachi, T. J.; Kariv-Miller, E. *J. Am. Chem. Soc.* **1988**, *110*, 3622.
62. Amatore, C.; Jutand, A. *Acta Chem. Scand.* **1990**, *44*, 755.
63. Marzouk, H.; Rollin, Y.; Folest, J. C.; Nédélec, J. Y.; Périchon, J. *J. Organometal. Chem.* **1989**, *369*, C47.
64. Garnier, L.; Rollin, Y.; Périchon, J. *J. Organometal. Chem.* **1989**, *367*, 347.
65. Labbé, E.; Duñach, E.; Périchon, J. *J. Organometal. Chem.* **1988**, *353*, C51.
66. Duñach, E.; Périchon, J. *J. Organometal. Chem.* **1988**, *352*, 239.
67. Dérien, S.; Clinet, J.-C.; Duñach, E.; Périchon, J. *J. Organometal. Chem.* **1992**, *424*, 213.
68. Dérien, S.; Clinet, J.-C.; Duñach, E.; Périchon, J. *J. Org. Chem.* **1993**, *58*, 2578.
69. Moutet, J.-C. *Org. Prep. Proc. Int.* **1992**, *24*, 309.
70. Ozaki, S.; Horiguchi, I.; Matsushita, H.; Ohmori, H. *Tetrahedron Lett.* **1994**, *35*, 725.
71. Kasiwagi, Y.; Ono, H.; Osa, T. *Chem. Lett.* **1993**, 81.
72. Kasiwagi, Y.; Ono, H.; Osa, T. *Chem. Lett.* **1993**, 257.
73. Inokuchi, T.; Matsumoto, S.; Fukushima, M.; Torii, S. *Bull. Chem. Soc. Jpn.* **1991**, *64*, 796.
74. Inokuchi, T.; Matsumoto, S.; Torii, S. *J. Org. Chem.* **1991**, *56*, 2416.
75. Navarro, M.; De Giovani, W. F.; Romero, J. R. *Tetrahedron* **1991**, *47*, 851.
76. Konno, A.; Fukui, K.; Fuchigami, T.; Nonaka, T. *Tetrahedron* **1991**, *47*, 887.
77. Palmisano, G.; Danieli, B.; Santagostino, M.; Vodopivec, B.; Fiori, G. *Tetrahedron Lett.* **1992**, *33*, 7779.
78. Matsuura, I.; Ueda, T.; Murakami, N. Nagai, S.-I.; Sakakibara, J.; Kurono, Y.; Hatano, K. *J. Chem. Soc., Chem. Commun.* **1992**, 828.
79. Nikishin, G. I. Elinson, M. N.; Makhova, I. V. *Tetrahedron* **1991**, *47*, 895.

80. Ginzel, K.-D.; Brungs, P.; Steckhan, E. *Tetrahedron* **1989**, *45*, 1691.
81. Hartstock, F. W.; McMahon, L. B.; Tell, I. P. *Tetrahedron Lett.* **1993**, *34*, 8067.
82. Bäckvall, J.-E.; Gogoll, A. *J. Chem. Soc., Chem. Commun.* **1987**, 1236.
83. Madurro, J. M.; Chiericato, G., Jr.; De Giovani, W. F.; Romero, J. R. *Tetrahedron Lett.* **1988**, *29*, 765.
84. Carrijo, R. M. C.; Romero, J. R. *Syn. Comm.* **1994**, *24*, 433.
85. Morita, M.; Masatan, T.; Matsuda, Y. *Bull. Chem. Soc. Jpn.* **1993**, *66*, 2647.
86. Nishihara, H.; Pressprich, K.; Murray, R. W.; Collman, J. P. *Inorg. Chem.* **1990**, *29*, 1000.
87. Horwitz, C. P.; Creager, S. E.; Murray, R. W. *Inorg. Chem.* **1990**, *29*, 1006.
88. Ojima, F.; Kobayashi, N.; Osa, T. *Bull. Chem. Soc. Jpn.* **1990**, *63*, 1374.
89. Leduc, P.; Battioni, P. Bartoli, J. F.; Mansuy, D. *Tetrahedron Lett.* **1988**, *29*, 205.
90. Bedioui, F.; Gutiérrez Granados, S.; Devynck, J.; Bied-Charreton, C. *New J. Chem.* **1991**, *15*, 939.
91. Coleman, J. P.; Hallcher, R. C.; McKackins, D. E.; Rogers, T. E.; Wagenknecht, J. H. *Tetrahedron* **1991**, *47*, 809.
92. Shundo, R.; Nishiguchi, I.; Matsubara, Y.; Hirashima, T. *Chem. Lett.* **1990**, 2285.
93. Shundo, R.; Nishiguchi, I.; Matsubara, Y.; Toyoshima, M.; Hirashima, T. *Chem. Lett.* **1991**, 185.
94. Shundo, R.; Nishiguchi, I.; Matsubara, Y.; Hirashima, T. *Chem. Lett.* **1991**, 235.
95. Shundo, R.; Nishiguchi, I.; Matsubara, Y.; Hirashima, T. *Tetrahedron* **1991**, *47*, 831.
96. Nédélec, J. Y.; Nohair, K. *Synlett.* **1991**, 659.
97. Giordano, C.; Belli, A.; Casagrande, F.; Guglielmetti, G.; Citterio, A. *J. Org. Chem.* **1981**, *46*, 3149 and references cited therein.
98. Fristad, W. E.; Fry, M. A.; Klang, J. A. *J. Org. Chem.* **1983**, *48*, 3575 and references cited therein.
99. Hansen, E. W.; Ruoff, P. *J. Phys. Chem.* **1989**, *93*, 264.
100. Zhang, Y.-Z.; Field, R. J. *J. Phys. Chem.* **1990**, *94*, 7154.
101. Ou, C.-C.; Jwo, J.-J. *Int. J. Chem. Kin.* **1991**, *23*, 137.
102. Orban, M.; Lengyel, I.; Epstein, I. R. *J. Am. Chem. Soc.* **1991**, *113*, 1978.
103. Berlett, B. S.; Chock, P. B.; Yim, M. B. *Proc. Nat. Acad. Sci.* **1990**, *87*, 384, 389 and 394.
104. Sarneski, J. E.; Michos, D.; Thorp, M. D.; Poon, T. Blewitt, J.; Brudvig, G. W.; Crabtree, R. H. *Tetrahedron Lett.* **1991**, *32*, 1153.
105. Allegretti, M.; D'annibale, A.; Trogolo, C. *Tetrahedron* **1993**, *49*, 10705.

RECEIVED August 4, 1994

Chapter 8

Supercritical Carbon Dioxide as a Medium for Conducting Free-Radical Reactions

James M. Tanko, Joseph F. Blackert, and Mitra Sadeghipour

Department of Chemistry, Virginia Polytechnic Institute and State University, Blacksburg, VA 24061-0212

Supercritical carbon dioxide (SC-CO_2) is found to be an excellent solvent for free radical brominations. Reaction yields, times, and selectivities are analogous to what is observed in conventional organic solvents (CCl_4, CFC's, and benzene). SC-CO_2 thus appears to be a suitable, "environmentally-benign" alternative solvent for free radical reactions.

There is an increased awareness of the need to reduce or eliminate toxic chemical waste and/or by-products which arise in the course of chemical synthesis and manufacture (*1-3*). A fundamental change in the philosophy of synthetic design is needed wherein the priority is placed upon health and environmental impact rather than just the efficiency (%-yield) associated with a chemical transformation. Certainly one aspect of synthesis in which dramatic advances in pollution prevention can be realized entails replacing many of the toxic, environmentally-threatening solvents utilized in most chemical processes with non-toxic, "environmentally-benign" alternatives. The challenge from the chemical perspective is to identify suitable alternatives. Our contribution to this effort has involved examining the potential use of supercritical carbon dioxide (SC-CO_2) as a solvent for free radical reactions. In this paper, we report our results pertaining to the free radical bromination of alkylaromatics in SC-CO_2 (*4*).

Generally, solvents suitable for free radical reactions do not possess reactive functionalities (e.g., abstractable hydrogens and reactive double bonds). Unfortunately, many of the solvents which meet this criteria are either carcinogenic (e.g., benzene) or damaging to the environment. Chlorofluorocarbons and carbon tetrachloride illustrate this latter problem as these materials are believed to be major culprits in the depletion of the earth's ozone layer (*5*).

Ozone in the earth's atmosphere shields the surface from damaging ultraviolet irradiation in the 280 - 320 nm region (equation 1). Nature maintains a delicate balance between this natural process which depletes ozone, and other natural processes which produce ozone. Freons and CCl_4 are believed to destroy this delicate balance. Chlorine atoms generated from these compounds (equation 2) introduce an unnatural mechanism for ozone depletion (5).

$$O_3 + h\nu \rightarrow O_2 + O \qquad (1)$$
$$CCl_2F_2 + h\nu \rightarrow {}^\bullet CClF_2 + Cl^\bullet \qquad (2)$$

At high altitudes (40 km), a process catalytic in Cl• is believed to operate (Scheme 1), while at lower altitudes where high concentrations of atomic oxygen are not available, a slightly different mechanism is suggested (Scheme 2). This latter process is believed to be responsible for the seasonal hole in the ozone layer over Antarctica (5).

Scheme 1

$$Cl^\bullet + O_3 \rightarrow ClO^\bullet + O_2$$
$$ClO^\bullet + O \rightarrow Cl^\bullet + O_2$$

$$O_3 + O \rightarrow 2\,O_2 \quad \textit{(overall)}$$

Scheme 2

$$2\,Cl^\bullet + 2\,O_3 \rightarrow 2\,ClO^\bullet + 2\,O_2$$
$$2\,ClO^\bullet \rightarrow ClOOCl$$
$$ClOOCl + h\nu \rightarrow Cl^\bullet + ClOO^\bullet$$
$$ClOO^\bullet \rightarrow Cl^\bullet + O_2$$

$$2\,O_3 + h\nu \rightarrow 3\,O_2 \quad \textit{(overall)}$$

To solve this problem, in accordance with international agreement (Montreal Protocol, 1987; London Amendment, 1990) the production of CFC's and CCl_4 is to be phased out in 1995. This "solution" of course generates a *new* problem: The chemical industry must find suitable "environmentally benign" alternative solvents for extractions, separations, processing and *synthesis*.

SC-CO_2 as a Reaction Solvent. The supercritical state is achieved when a substance is taken above its critical temperature and pressure. The bulk properties of a supercritical fluid are intermediate between those of a gas and a liquid. Because of the unique properties of supercritical fluids, analytical methods based upon their use are becoming increasingly popular on both a laboratory and industrial scale (chromatography, extractions, etc...) (6).

In a similar vein, there are several potential advantages which may be realized with the use of SC-CO_2 as a solvent for chemical reactions:

- Important solvent properties of SC-CO_2 (e.g., dielectric constant, solubility parameter, viscosity, density) can be altered via manipulation of temperature and pressure (7,8). This unique property of a supercritical fluid could be exploited to control the behavior (e.g., kinetics and selectivity) of some chemical processes.

- The supercritical state of CO_2 is relatively easily attained (T_c = 31 °C, P_c = 74 bar) (7). This feature of SC-CO_2 implies that the "cost" (e.g., energy, apparatus, and materials) will not be prohibitive with the conversion to SC-CO_2.

- Finally, CO_2 is non-toxic and "environmentally benign."

Another factor to be considered in the use of SC-CO_2 as a reaction solvent deals with the possible effect of pressure on reaction rate (and selectivity) (9). For the hypothetical reaction A + B → C, the effect of pressure on reaction rate is expressed by equation 3, where k is the rate constant for the reaction and $\Delta V^{\neq}$ (the volume of activation) is the difference in molar volume between the transition state and reactants ($\Delta V^{\neq} = V_{ts} - V_A - V_B$) (10). Reactions exhibiting negative volumes of activation imply a transition state which is smaller and more compact than the reactants (e.g., an associative process). Similarly, a positive volume of activation implies expansion as the reactants progress to the transition state (e.g., a dissociative process).

$$[\delta \ln k / \delta P]_T = -\Delta V_{act} / RT \qquad (3)$$

The Diels-Alder reaction is characterized by substantially negative volumes of activation (ca. -20 → -50 cm^3/mol, depending on the substrates) (11), and has been extensively studied in SC-CO_2 (12-16). At high CO_2 pressures, the measured volume of activation is analogous to that observed in conventional solvents. However, at pressures just above the critical pressure of CO_2, there is a surprisingly large variation of k with pressure ($\Delta V^{\neq}$ is on the order of -500 cm^3/mol) (12-16). Similar variations of rate over a narrow range of pressures just above the critical pressure have also been noted for other reactions in supercritical fluid solvent (17). This dramatic variation of rate with pressure just above the critcal point is attributed to a phenomenon referred to as "clustering" in which solvent molecules are believed to aggregate (cluster) about the solute (12-17).

Free Radical Reactions in SC-CO_2. Literature Precedent. The issue as to whether CO_2 is inert to free radicals must be addressed. Since CO_2 formally possesses a carbon-oxygen double bond, there exists a possibility of radical addition to CO_2 (equation 4). In conventional solvents, the equilibrium depicted in equation 4 favors reactants. (Decarboxylation of carboxyl radicals is facile, e.g., the Kolbe electrolysis) *(18)*. What effect the high pressure associated with the $SC-CO_2$ medium might have on this equilibrium is unclear.

$$R\cdot + CO_2 \rightleftharpoons RCO_2\cdot \qquad (4)$$

Literature precedent suggests that $SC-CO_2$ is inert to *stabilized* carbon-centered radicals (e.g., benzyl). For example, McHugh reported the autooxidation of cumene in $SC-CO_2$ via the free radical chain process outlined in Scheme 3 *(19)*. This report is significant because it demonstrates for the first time that it is possible to conduct free radical chain reactions in $SC-CO_2$.

Scheme 3

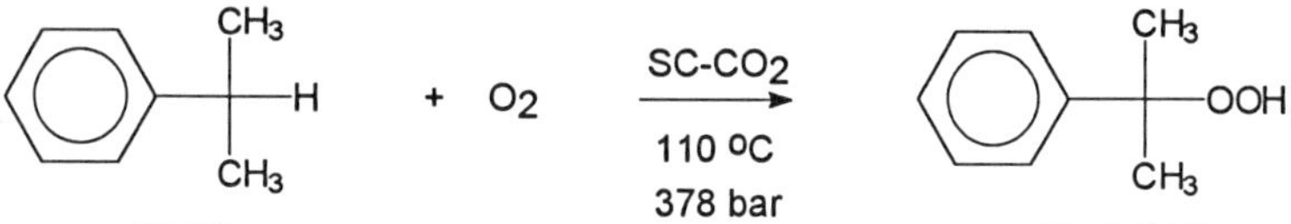

PhC(CH$_3$)$_2$H + O$_2$ $\xrightarrow{SC-CO_2, 110\ ^\circ C, 378\ bar}$ PhC(CH$_3$)$_2$OOH

(R-H) (R-OOH)

$$R\cdot + O_2 \longrightarrow ROO\cdot$$

$$ROO\cdot + R-H \longrightarrow ROOH + R\cdot$$

Fox and Johnston *(20)* reported that benzyl radicals generated photolytically from dibenzyl ketone dimerize yielding the corresponding bibenzyls (Scheme 4). An intriguing aspect of this report is that for unsymmetrical dibenzyl ketones, the resulting bibenzyls (A-A, A-B, and B-B) were formed in a statistical 1:2:1 ratio, suggesting that cage effects were unimportant (Scheme 4). The lifetime of the geminate A•/B• caged-pair is short because of the low viscosity of $SC-CO_2$ *(20)*.

DeSimone and co-workers have examined the kinetics of decomposition of the free radical initiator 2,2'-azobis(isobutyronitrile) (AIBN) in $SC-CO_2$ and have also demonstrated the possible synthetic utility of $SC-CO_2$ via the free radical polymerization of 1,1-dihydroperfluorooctyl acrylate (FOA, equation 5) *(21,22)*.

Objectives. The goal of our work is to examine and assess the feasiblity of using $SC-CO_2$ as a solvent for free radical chain reactions for synthetic purposes. Because of the "tunable" solvent properties of $SC-CO_2$, this medium offers unique advantages over conventional solvents from a *chemical* perspective (in addition to its obvious advantages from an environmental perspective). As noted earlier, solvent properties of $SC-CO_2$ such as viscosity and polarity vary as a function of temperature and pressure. It is thus conceivable that for reactions sensitive to the effects of solvent viscosity, solvent polarity, or pressure that reactivity/selectivity could be "dialed-up"

Scheme 4

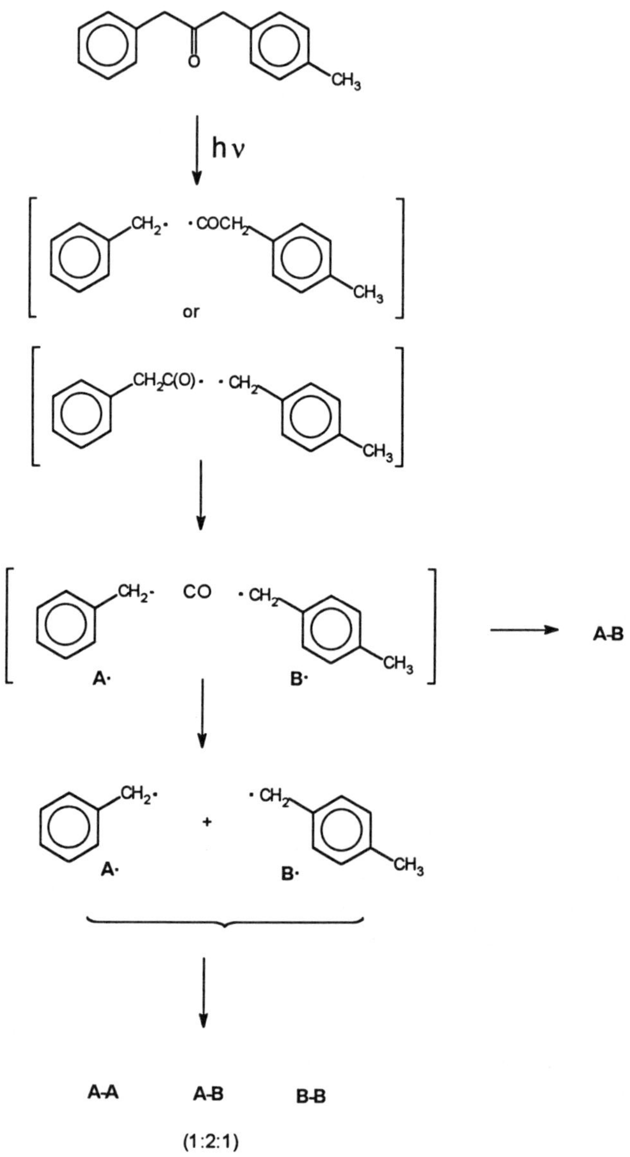

A·A   A·B   B·B
(1:2:1)

$$CH_2=CHCO_2CH_2C_7F_{15} \xrightarrow[\substack{59.4\ °C \\ SC\text{-}CO_2 \\ 207\ bar}]{\substack{AIBN \\ 48\ hours}} \ \ \{CH_2CH\}_n\ |\ CO_2CH_2C_7F_{15} \quad (5)$$

by the appropriate manipulation of temperature and pressure. This type of external control over a chemical process is not possible in conventional solvents.

This project began with our study of the free radical bromination of alkylaromatics in SC-CO_2 with the purpose of addressing the following issues:

- Is SC-CO_2 inert towards free radicals?

- Are reaction yields compromised by the use of SC-CO_2?

- How is reactivity/selectivity affected in SC-CO_2 medium?

Free Radical Bromination of Alkylaromatics. Reaction of molecular bromine with an alkylaromatic (e.g., toluene) in a suitable solvent initiated either photochemically (hv) or thermally with a free radical initiator produces the corresponding benzylic bromide in nearly quantitative yield. The mechanism of the reaction, illustrated in Scheme 5 for the reaction of toluene, involves a free radical chain process in which: a) hydrogen abstraction by atomic bromine (Br•) generates benzyl radical (PhCH$_2$•) and HBr, and b) trapping of PhCH$_2$• by molecular bromine produces benzyl bromide and regenerates Br•. Long chain lengths are observed for these reactions, and the selectivity associated with hydrogen abstractions by Br• is very high (generally the weakest C-H bond in the substrate is broken) (23).

Experimental

Precautions. Because of the high pressures associated with supercritical fluids, special attention was paid to ensuring that the pressures utilized in these experiments did not exceed the specifications of our apparatus (estimated to be ca. 15,000 psi). Although we experienced no accidents in these experiments, appropriate cautions were always taken (in addition to the liberal use of common sense) because of the potential hazards associated with high pressure work.

General Considerations. Gas chromatographic analyses were performed on a Hewlett-Packard HP 5890A equipped with FID detectors and an HP 3393A reporting integrator. Analyses were accomplished on a 30 m SE-30 column (0.25 mm diameter). Solvents were obtained from Aldrich and distilled prior to use. Toluene and ethyl benzene were obtained from Aldrich, distilled, and stored over 4 Å molecular sieves under argon prior to use.

Scheme 5

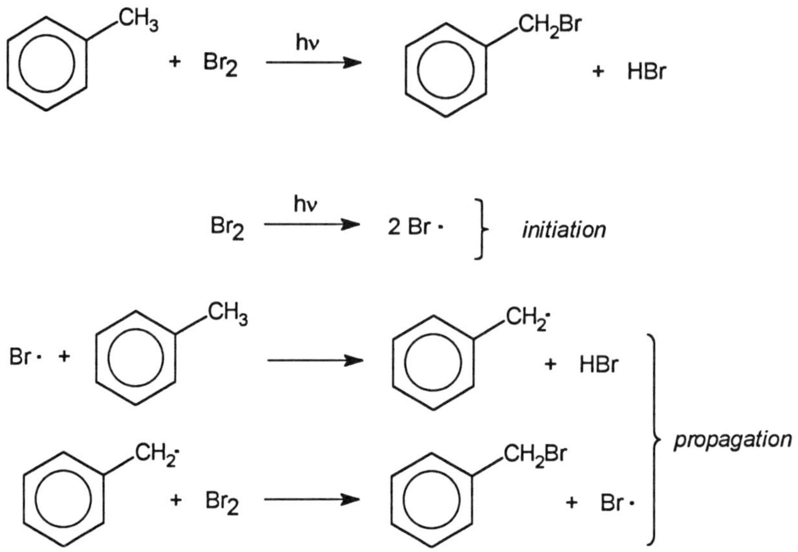

$C_6H_5CH_3 + Br_2 \xrightarrow{h\nu} C_6H_5CH_2Br + HBr$

$Br_2 \xrightarrow{h\nu} 2\, Br\cdot \quad \}$ *initiation*

$Br\cdot + C_6H_5CH_3 \longrightarrow C_6H_5CH_2\cdot + HBr$

$C_6H_5CH_2\cdot + Br_2 \longrightarrow C_6H_5CH_2Br + Br\cdot \quad \}$ *propagation*

General Considerations (SC-CO$_2$ experiments). The high pressure reactor utilized for the supercritical CO$_2$ experiments was constructed in house, and consisted of a sapphire window (for irradiation) and a magnetic stir bar. Because of the corrosive nature of Br$_2$ (and in an effort to minimize possible by-products arising from Fe-catalyzed electrophilic aromatic substitution) the reactor was fabricated from Hastelloy C-276 alloy, which is corrosion-resistant. (A schematic of the reactor is provided in Figure 1). The complete apparatus for the supercritical experiments is outlined in Figure 2: Temperature control was provided by a Lauda K 6 oil circulating bath via a stainless steel heat exchange coil. Pressures were achieved utilizing a high pressure piston (High Pressure Equipment Co.), measured with a pressure transducer (Sensotec Model 8671-01 transducer and Model GM readout). SFC-grade CO$_2$ was obtained from Scott Specialty Gases.

Free radical brominations in SC-CO$_2$. A sealed glass ampule containing Br$_2$ (Freeze-pump-thaw [FPT] degassed [freezing to -196 °C, pumping to < 20 milliTorr, and thawing at room temperature]) was introduced into the high pressure reactor (ca. 25 mL capacity). The alkylaromatic, previously degassed by bubbling argon, was introduced into the reactor via syringe under argon backflush. The reactor was sealed and pressurized to 50 psi with CO$_2$ and vented three times to remove oxygen. The reactor was then pressurized with CO$_2$ to the desired pressure (during which time the Br$_2$ ampule usually ruptured), and equilibrated at 40 °C for 5 - 10 min. Br$_2$ color was evident in solution until exposure to UV light (supplied by a 400 W medium pressure mercury arc lamp). Once irradiated, Br$_2$ color dissipated in less than 5 min. The cell was depressurized by bubbling the contents through hexane. Once the reactor had returned to ambient pressure, the cell was washed

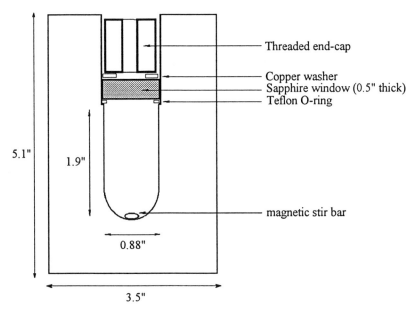

Figure 1. Cross-section schematic of the reactor used for reactions in SC-CO_2 solvent

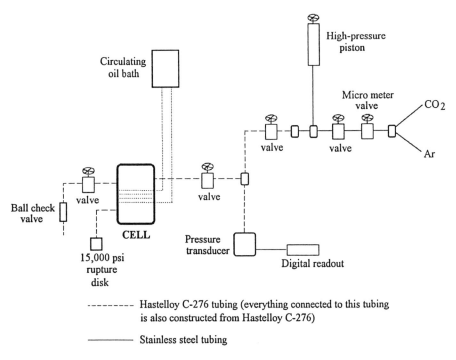

Figure 2. Schematic diagram of the apparatus used to generate SC-CO_2

repeatedly with hexane and the combined hexane washings were analyzed by gas chromatography vs an appropriate internal standard.

Free radical brominations in conventional solvent. An appropriate quantity of alkylaromatic and 10 mL solvent were placed in a 30-mL Pyrex pressure tube (equipped with an O-ringed Teflon needle valve and Teflon-coated magnetic stir bar). An appropriate HBr scavenger (1,2-epoxybutane) was added, and the solution was degassed 4x by the freeze-pump-thaw (FPT) method. Bromine was FPT degassed and distilled (via vacuum line) into the reaction mixture at -196 °C. The pressure tube was sealed and the reaction mixture allowed to equilibrate at the desired reaction temperature in total darkness. The reaction mixture was irradiated with a 400 W medium-pressure mercury arc lamp at a distance of 2 ft through two Pyrex layers. Complete discharge of Br_2 occurred in less than 5 min. Afterward, the solution was analyzed by GLC vs an appropriate internal standard.

Results and Discussion

Bromination of Alkylaromatics in SC-CO_2. Product Yields. The bromination of toluene in SC-CO_2 proceeded smoothly as indicated in equation 6. The major reaction product, formed in ≥70% yield, was benzyl bromide accompanied by a small amount (10 - 20%) of *p*-bromotoluene (resulting from a competing electrophilic aromatic substitution process).

PhCH$_3$ + Br$_2$ $\xrightarrow[\substack{SC-CO_2 \\ 252 \text{ bar} \\ K_2CO_3 \\ 5 \text{ min.}}]{\substack{h\nu \\ 40\ ^\circ C}}$ PhCH$_2$Br + *p*-BrC$_6$H$_4$CH$_3$ (6)

| 10 | : | 1 | 83% | 17% |
| 2 | : | 1 | 74% | 11% |

Similarly, bromination of ethylbenzene in SC-CO_2 yielded in 1-bromo-1-phenylethane in nearly quantitative yield (equation 7).

Competitive Bromination of Ethylbenzene and Toluene in SC-CO_2. Via direct competition (Scheme 6), the selectivity (relative reactivity of the "secondary" hydrogens of ethylbenzene and the "primary" hydrogens of toluene on a per-hydrogen basis), $r(2^\circ/1^\circ)$, in both SC-CO_2 and conventional solvent was calculated from product yields: $r(2^\circ/1^\circ) = (k_{Et}/k_{Me}) \times (3/2) = $ (Yield $C_6H_5CHBrCH_3$/Yield $C_6H_5CH_3$) × ($[PhCH_3]_i/[PhCH_2CH_3]_i$) × (3/2). (For all experiments, $[Br_2]_i << [PhCH_3]_i, [PhCH_2CH_3]_i$). The results are summarized in Table I.

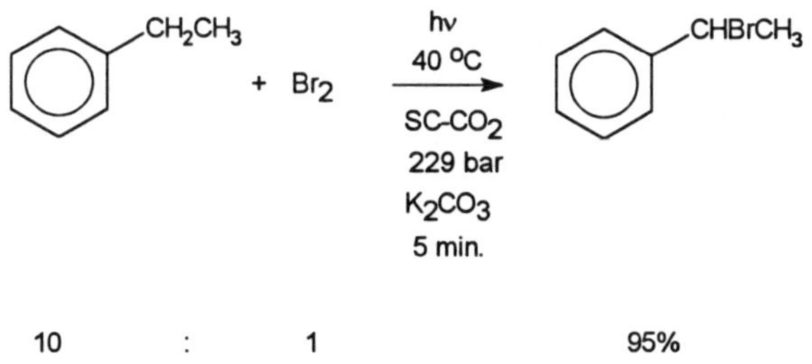

PhCH₂CH₃ + Br₂ →(hv, 40 °C, SC-CO₂, 229 bar, K₂CO₃, 5 min) PhCHBrCH₃   (7)

10   :   1   95%

Scheme 6

Br• →(k_Et, PhCH₂CH₃) PhĊHCH₃ →(Br₂) PhCHBrCH₃ **A**

Br• →(k_Me, PhCH₃) PhCH₂• →(Br₂) PhCH₂Br **B**

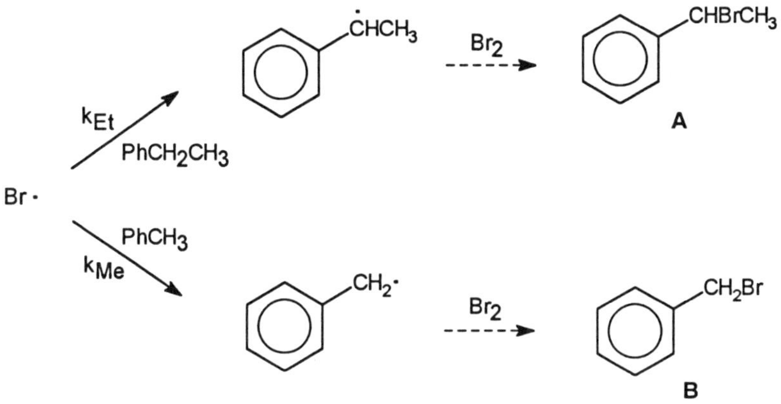

Table I. Summary of Toluene/Ethylbenzene Competitions for Br• in SC-CO$_2$ at 40°C

Pressure, bar	$r(2^o/1^o)$
75	30 ± 2
79	31 ± 2
111	28 ± 1
119	30 ± 1
247	25 ± 2
339	32 ± 1
423	30 ± 1

Within experimental error, reaction selectivity does not vary with pressure. The invariance of r(2°/1°) with pressure can be explained because the difference in the volume of activation for hydrogen abstraction from ethylbenzene vs toluene by Br• is small ($\Delta\Delta V_{act}$ = -4.8 cm^3/mol) (*11*). On this basis, over the range of pressures examined in this work (ca. 75 - 400 bar), the relative reactivity would vary by less than 10% (i.e., the anticipated pressure effect is of the same magnitude as experimental error). Moreover, even at pressures just above the supercritical pressure of CO_2 we do not find any evidence of altered selectivity which could be attributed to CO_2 solvent clustering.

The data in Table I also provide convincing evidence that Br• was the chain carrier in these reactions as opposed to a Br•/CO_2 complex or adduct. (Chlorine atom selectivities have been found to vary when free radical chlorinations are conducted in CS_2 solvent because of the formation of a Cl•/CS_2 complex) (*24,25*). The selectivity in SC-CO_2 is nearly identical to that observed in conventional organic solvents (Table II). The fact that the selectivity is nearly the same in SC-CO_2 and conventional organic solvent strongly implicates free Br• as the chain carrier in SC-CO_2.

Table II. Summary of Toluene/Ethylbenzene Competitions for Br• in Conventional Organic Solvents

solvent	r(2°/1°), 17 °C	r(2°/1°), 40 °C
CCl_4	42±1	35±1
Freon 113	40±1	34±1
CH_2Cl_2	25±1	29±1

Ziegler (NBS) bromination of toluene and ethylbenzene in SC-CO_2. The Ziegler bromination is an important method for the synthesis of allylic and benzylic bromides (equation 8) (*26*). Critical features to the success of this reaction include the use of N-bromosuccinimide (NBS) as the brominating agent and CCl_4 as reaction solvent. The advantage of the Ziegler method over direct bromination with Br_2 is that competing electrophilic processes are virtually eliminated. An appreciation as to why electrophilic processes are minimized can be derived from an analysis of the reaction mechanism.

The mechanism of the Ziegler bromination is depicted in Scheme 7 (*27*). The reason that CCl_4 is the ideal solvent for this reaction arises from the fact that NBS is insoluble in CCl_4. As such, there is insufficient NBS in solution to allow the propagation of a succinimidyl radical chain. The role of NBS in this reaction is to maintain a low, steady-state concentration of Br_2 in solution. Because of the very low Br_2 concentrations produced under the reaction conditions, the rates of competing electrophilic processes are very slow, and hence, are virtually eliminated (27).

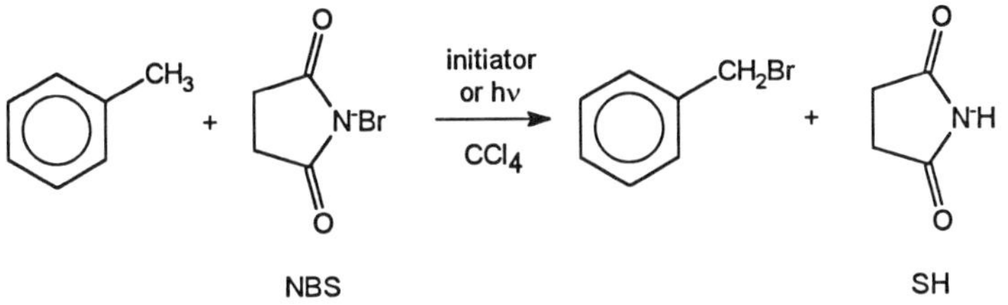

$$\text{toluene} + \text{NBS} \xrightarrow[CCl_4]{\text{initiator or } h\nu} \text{benzyl bromide} + \text{SH} \quad (8)$$

Scheme 7

$$Br\cdot + \text{toluene} \longrightarrow \text{benzyl radical} + HBr$$

$$\text{benzyl radical} + Br_2 \longrightarrow \text{benzyl bromide} + Br\cdot$$

$$HBr + \text{NBS} \longrightarrow \text{SH} + Br_2$$

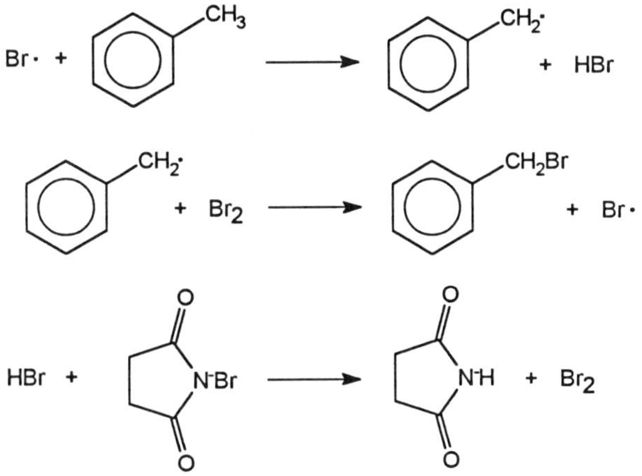

Reaction of toluene, NBS, and 10 mol-% AIBN (initiator) in SC-CO_2 (170 bar, 40 °C) irradiated for four hours yielded benzyl bromide in quantitative yield with no detectable amount of *p*-bromotoluene (equation 9). As is the case for reactions conducted in CCl_4, NBS was observed to be insoluble in SC-CO_2 (i.e, the reaction mixture was *heterogeneous*).

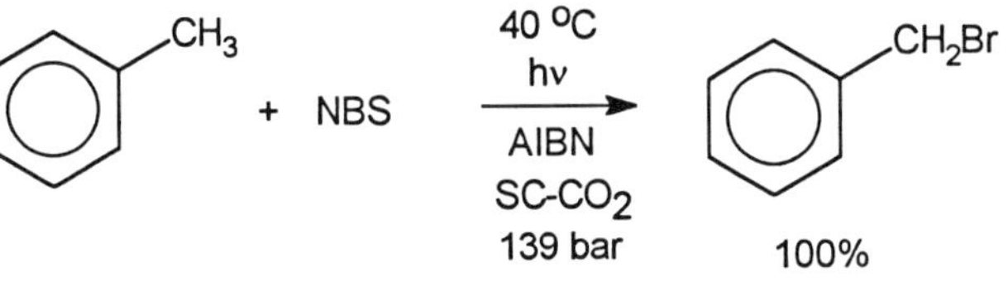

$$\text{toluene} + \text{NBS} \xrightarrow[\substack{AIBN \\ SC-CO_2 \\ 139 \text{ bar}}]{\substack{40 \text{ °C} \\ h\nu}} \text{benzyl bromide} \quad 100\% \quad (9)$$

A competition experiment pitting ethylbenzene vs toluene in SC-CO_2 (40 °C, 342 bar) yielded a relative reactivity, $r(2°/1°) = 29 \pm 2$. This result is identical to that observed with direct bromination with molecular bromine (Table I) and suggests that Br• is the chain carrier for the Ziegler bromination in SC-CO_2. Consequently, there does not appear to be a change of mechanism with the conversion from CCl_4 to SC-CO_2 as solvent for the Ziegler bromination.

Free radical bromination of cyclopropylbenzene. In order to evaluate further the properties of SC-CO_2 as a solvent for free radical brominations, we examined the free radical bromination of cyclopropylbenzene in this medium (equation 10).

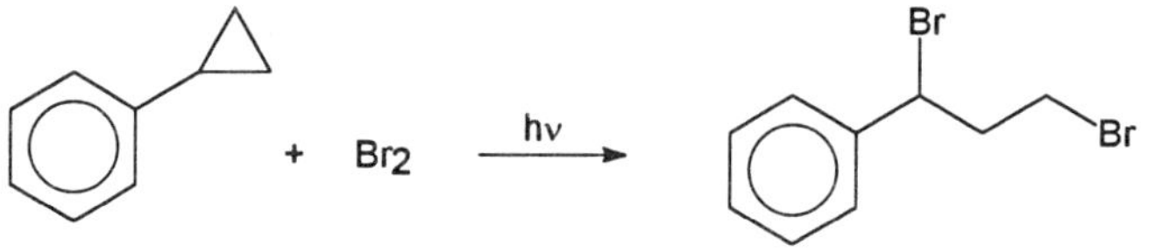

$$PhCH(C_3H_5) + Br_2 \xrightarrow{h\nu} Ph-CHBr-CH_2-CH_2-Br \qquad (10)$$

Unlike other alkylaromatics which yield substitution products in their reaction with Br_2, cyclopropylbenzene yields an addition product (1,3-dibromo-1-phenylpropane) arising from the S_H2 process outlined in Scheme 8 (28-30). As with conventional organic solvents, this reaction proceeds in quantitative yield in SC-CO_2.

Scheme 8

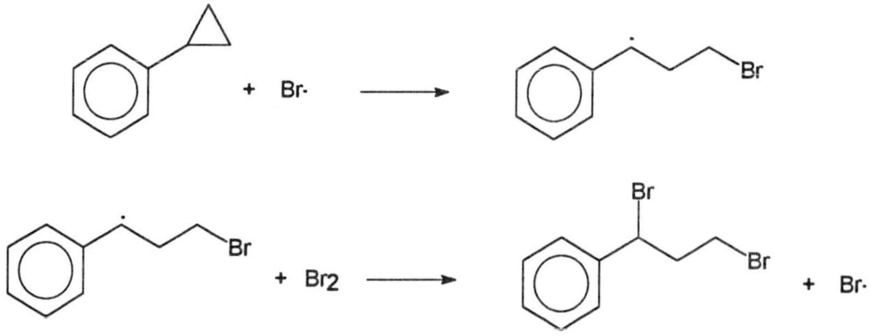

$$PhC_3H_5 + Br• \longrightarrow Ph-CH•-CH_2-CH_2-Br$$

$$Ph-CH•-CH_2-CH_2-Br + Br_2 \longrightarrow Ph-CHBr-CH_2-CH_2-Br + Br•$$

The relative rate constants for reaction of Br• with toluene (k_H) and cyclopropylbenzene (k_C) in SC-CO_2 were assessed by direct competition (Scheme 9). The rate constant ratio k_C/k_H in SC-CO_2 (40 °C, 1,300 psi) was 1.2 ± 0.2 and is nearly identical to that found in CCl_4 at the same temperature (1.3 ± 0.1). These observations also support the hypothesis that Br• properties (i.e., reactivity/selectivtiy) are unaffected by complexation to CO_2 solvent. Earlier work has shown that k_C/k_H varies as a function of solvent internal pressure (31). We are presently examining the effect of CO_2 pressure on k_C/k_H.

Scheme 9

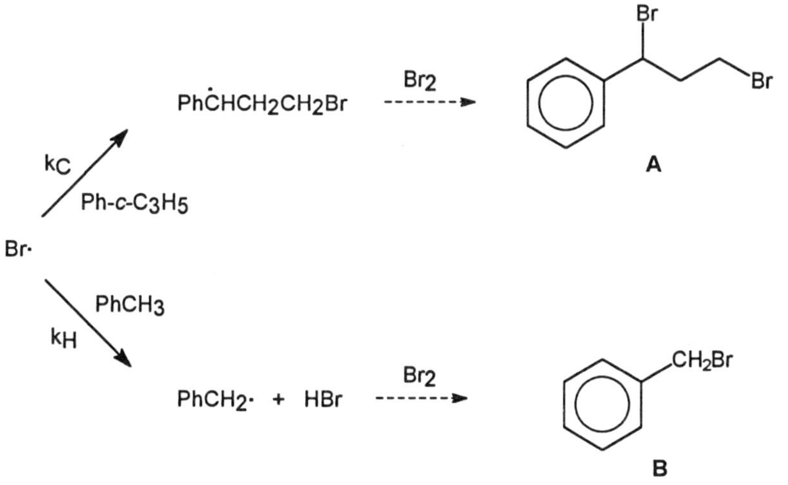

PhĊHCH₂CH₂Br →(Br₂) **A**: PhCH(Br)CH₂CH₂Br

Br• →(kC, Ph-c-C₃H₅) PhĊHCH₂CH₂Br

Br• →(kH, PhCH₃) PhCH₂• + HBr

PhCH₂• →(Br₂) **B**: PhCH₂Br

k_C/k_H = (Yield-A/Yield-B) × [PhCH₃]$_i$/[Ph-c-C₃H₅]$_i$

[Br₂]$_i$ < [PhCH₃]$_i$, [Ph-c-C₃H₅]$_i$

Conclusions

The results presented herein demonstrate that free radical brominations can be conducted effectively in SC-CO₂ as solvent. The high product yields and selectivities usually found for these reactions in conventional solvents are not compromised by the use of this non-toxic, less environmentally threatening medium. SC-CO₂ emerges as an excellent (and perhaps superior) solvent for free radical brominations.

In future work, we hope to:

• illustrate the utility of SC-CO₂ as a solvent for other types of free radical reactions

• demonstate that the tunable properties of SC-CO₂ may provide a means of "dialing-up" the reactivity/selectivity of radical intermediates.

In terms of the overall goal of pollution prevention, the results reported herein (and in literature citations) demonstrate that supercritical fluid solvents retain virtually all of the "chemical" advantages associated with conventional organic solvents. It seems inevitable that SC-CO₂ will replace many of the hazardous solvents currently employed for chemical synthesis and manufacture. While it is feasible that other less-hazardous, synthetic materials could be developed, by virtue of the fact that CO_2 is a *naturally-occurring* component of the atmosphere, its use is far less likely to result in unanticipated environmental damage than any synthetic alternative. Thus, the goals of "pollution prevention" and "environmentally-benign synthesis" appear to be realistic and attainable. The successful use of CO_2 as a solvent represents an important early step in the growth and development of "green chemistry."

Acknowledgments

Financial support from the U.S. Environmental Protection Agency, Office of Pollution Prevention and Toxics, Design for the Environment Program, the National Science Foundation (CHE 9113448), and the Thomas F. and Kate Miller Jeffress Memorial Trust Fund is gratefully acknowledged. We also thank Mr. Fred Blair (Chemistry/Physics Machine Shop) for his efforts in the construction of the high pressure reactors used in this work, and Prof. Joseph DeSimone (University of North Carolina, Chapel Hill) for his advice and assistance in setting up our $SC-CO_2$ system.

Literature Cited

1. Amato, I. *Science* **1993**, *259*, 1538.
2. Wedin, R. E. *Todays Chemist at Work* **1993**, *2*, 16.
3. Illman, D. *Chem. and Eng. News* **1993**, *71* (May 29, 1993), 5.
4. A preliminary account of this work has appeared: Tanko, J. M.; Blackert, J. F. *Science* **1994**, *263*, 203.
5. Zurer, P. S. *Chem. and Eng. News* **1993**, *71* (May 24, 1993), 8.
6. *Supercritical Fluid Science and Technology;* Johnston, K. P.; Penninger, J. M. L., Eds.; ACS Symposium Series 406; American Chemical Soceity: Washington, DC, 1989.
7. McHugh, M. A.; Krukonis, V. J. *Supercritical Fluid Extraction Principles and Practice*; Butterworths: Boston, MA, 1986.
8. *Supercritical Fluid Technology. Reviews in Modern Theory and Applications*; Bruno, T. J.; Ely, J. F., Eds.; CRC Press: Boca Raton, FL, 1991.
9. For a discussion, see Sunwook, K; Johnston, K. P. In *Supercritical Fluids: Chemical and Engineering Principles and Applications*; Squires, T.G.; Paulaitis, M. E. Eds.; ACS Symposium Series 329; American Chemical Society: Washington, DC, 1987; pp. 42 - 55.
10. Moore, J. W.; Pearson, R. G. *Kinetics and Mechanism, 3rd Ed.*; Wiley: New York, NY, 1981; pp. 276 - 278.
11. Asano, T.; LeNoble, W. J. *Chem. Rev.* **1978**, *78*, 407.
12. Paulaitis, M. E.; Alexander, G. C. *Pure Appl. Chem.* **1987**, *59*, 61.
13. Kim, Sunwook; Johnston, K. P. *Chem. Eng. Comm.* **1988**, *63*, 49.
14. Isaacs, N. S.; Keating, N. *J. Chem. Soc. Chem. Commun.* **1992**, 876.
15. Ikushima, Y.; Ito, S; Asano; T.; Yokoyama, T.; Saito, N.; Hatakeda, K.; Goto, T. *J. Chem. Eng. Jpn.* **1990**, *23*, 96.
16. Ikushima, Y.; Saito, N.; Arai, M. *J. Chem. Phys.* **1992**, *96*, 2293.

17. Roberts, C. B.; Chateauneuf, J. E.; Brennecke, J. F. *J. Am. Chem. Soc.* **1992**, *114*, 8455.
18. For a discussion, see Leffler, J. E. *An Introduction to Free Radicals;* Wiley: New York, NY, 1993; p 153.
19. Suppes, G. J.; Occhiogrosso, R. N.; McHugh, M. A. *Ind. Eng. Chem. Res.* **1989**, *28*, 1152.
20. O'Shea, K. E.; Combes, J. R.; Fox, M. A.; Johnston, K. P. *Photochem. Photobiol.* **1991**, *54*, 571.
21. DeSimone, J. M.; Guan, Z.; Elsbernd, C. S. *Science* **1992**, *257*, 945.
22. Guan, Z.; Combes, J. R.; Menceloglu, Y. Z.; DeSimone, J. M. *Macromolecules* **1993**, *26*, 2663.
23. Poutsma, M. L. In *Free Radicals, Vol. II*; Kochi, J. K., Ed.; Wiley: New York, NY, 1973; pp. 159 - 229.
24. Russell, G. A. *J. Am. Chem. Soc.* **1958**, *80*, 4897.
25. Chateauneuf, J. E. *J. Am. Chem. Soc.* **1993**, *115*, 1915.
26. Ziegler, K.; Späth, A.; Schaaf, E; Schumann, W.;Winkelmann, E. *Liebigs Ann. Chem.* **1942**, *551*, 80.
27. Lüning, U.; Skell, P. S. *Tetrahedron* **1985**, *41*, 4189.
28. Maynes, G. G.; Applequist, D. E. *J. Am. Chem. Soc.* **1973**, *95*, 856.
29. Shea, K. J.; Skell, P. S. *J. Am. Chem. Soc.* **1973**, *95*, 6728.
30. Tanko, J. M.; Mas, R. H.; Suleman, N. K. *J. Am. Chem. Soc.* **1990**, *112*, 5557.
31. Tanko, J. M. Suleman, N. K.; Hulvey, G. A.; Park, A.; Powers, J. E. *J. Am. Chem. Soc.* **1993**, *115*, 4520.

RECEIVED August 4, 1994

Chapter 9

The University of California–Los Angeles Styrene Process

Orville L. Chapman

Department of Chemistry and Biochemistry,
University of California–Los Angeles, Los Angeles, CA 90024–1569

Styrene manufacture utilizes approximately 13 billion pounds of benzene each year. Existing technology alkylates benzene with ethylene and then dehydrogenates the ethylbenzene. Environmental considerations dictate that we should replace benzene, and the UCLA styrene process can. This process uses only mixed xylenes, which are more environmentally friendly than benzene. The UCLA styrene process converts equilibrium mixed xylenes, the cheapest aromatic source available, to styrene in a single high-temperature step. The mechanism of this remarkable oxidative rearrangement is complex, but the overall process is very simple.

The existing styrene process uses two starting materials, two steps, and two catalysts, and rests on roughly one hundred million dollars in research and development. The first step alkylates benzene with ethylene using an acid catalyst. The second step dehydrogenates ethyl benzene to styrene using a dehydrogenation catalyst. Both steps give high yields with very low production of by-products. Research and development and capital investment have long ago been recouped, and the process makes money. The only problem is a nagging fear that the chemical industry is going to have to do without benzene. In human livers, the P-450 enzyme system converts benzene to benzene epoxide, which is a carcinogen. Concern about the safety of benzene and the scale on which it is used--about 13 billion lbs/year in the United States--has fostered sporadic attempts to find an economically satisfactory replacement process.

9. CHAPMAN The UCLA Styrene Process

$$\text{C}_6\text{H}_6 + \text{H}_2\text{C}=\text{CH}_2 \xrightarrow{\text{Catalyst}} \text{PhCH}_2\text{CH}_3$$

$$\text{PhCH}_2\text{CH}_3 \xrightarrow{\text{Catalyst}} \text{PhCH}=\text{CH}_2$$

Alternatives to benzene are indeed few, and attention has focused primarily on toluene. Innes and Swift have summarized and critically reviewed the work with toluene (1). No really satisfactory alternative process exists. As strange as it may seem, no one has published work on xylenes as an alternative to benzene in styrene production even though the xylenes have eight carbons as does styrene.

$$\text{PhCH}_3 \xrightarrow{\text{PbO}} \text{HCPh}=\text{CHPh}$$

$$\text{PhCH}=\text{CHPh} + \text{H}_2\text{C}=\text{CH}_2 \xrightarrow{\text{Olefin Metathesis}} \text{PhCH}=\text{CH}_2$$

The UCLA styrene process derives from our investigation of the mechanism of the transformation of the isomeric diazomethyltoluenes to benzocyclobutane and styrene in low yield (2-5). An alternative process for a major industrial commodity chemical thus began with a purely academic investigation of reaction mechanism. The mechanism of these intriguing rearrangements begins with nitrogen expulsion giving the tolylmethylenes. The tolylmethylenes interconvert through the

$$\text{(3-CH}_3\text{-C}_6\text{H}_4\text{)CH}=\overset{+}{\text{N}}=\overset{-}{\text{N}} \xrightarrow{\text{Heat or Light}} \text{(3-CH}_3\text{-C}_6\text{H}_4\text{)}\ddot{\text{C}}\text{H} + \text{N}_2$$

Tolyldiazomethane   Tolylmethylene

methylcycloheptatetraenes ultimately yielding benzocyclobutane and styrene. Using low-temperature matrix-isolation spectroscopy with infrared, ultraviolet, and electron spin resonance spectroscopy as probes and guided by our earlier characterization of 1,2,4,6-cycloheptatetraene (6-7), we observed and characterized each of the intermediates in Scheme 1 (8-9). At the suggestion of industrial friends, we turned our attention to the possibility of making styrene the sole product of the

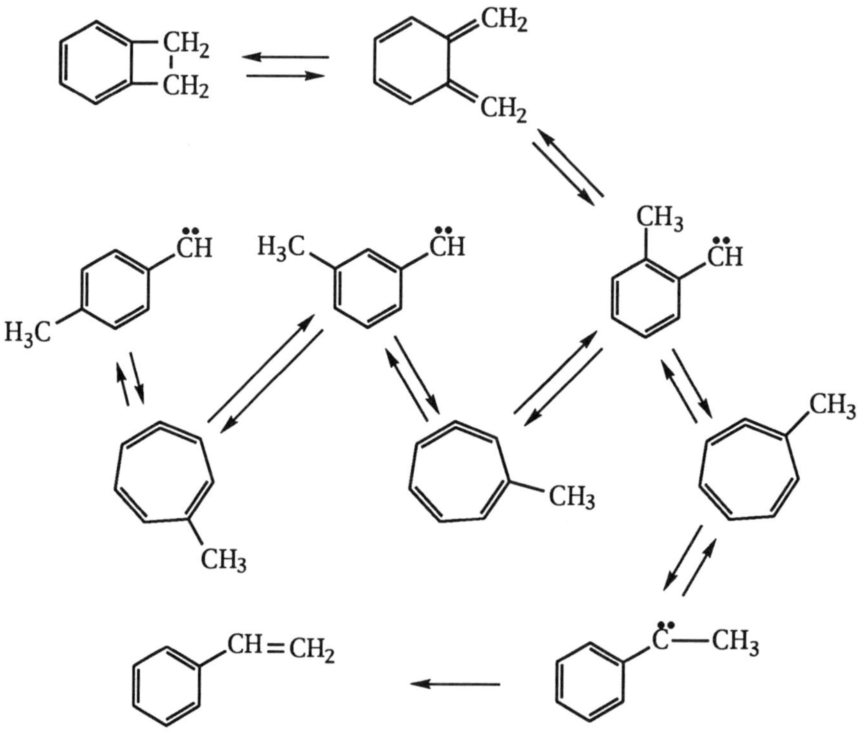

Scheme 1

process. We needed two things to make styrene. First, we had to convert benzocyclobutane to styrene, and second, we had to find an economically satisfactory route to the tolylmethylenes. In point of fact, the second problem proved to be the limiting factor.

The conversion of benzocyclobutane to styrene was, in fact, a solved problem (2-5). Baron and DeCamp had reported a 94% yield at 930 degrees Centigrade (3).

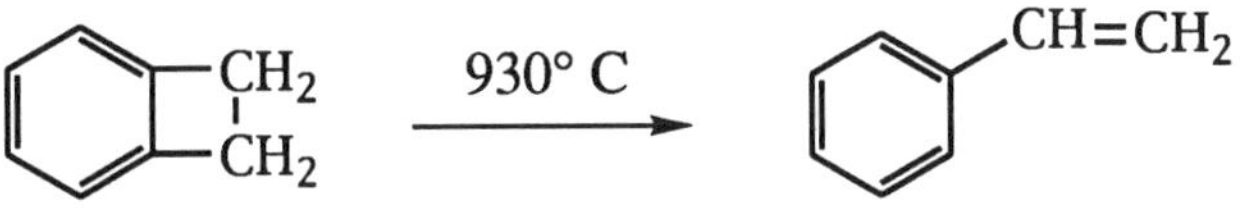

9. CHAPMAN The UCLA Styrene Process

$$\text{benzocyclobutane} \xrightarrow{930° C} \text{styrene}$$

We were interested in knowing whether the mechanism that we had elucidated at low temperature using photochemical interconversion applied also in the high-temperature domain. Scheme 2 shows two possible mechanisms, which carbon-13 labeling can distinguish. The standard diradical mechanism predicts equal labeling in the *alpha* and *beta* carbons of styrene; the mechanism based on our earlier studies predicts equal labeling in the *ortho* and *beta* positions of styrene.

Mechanism 1

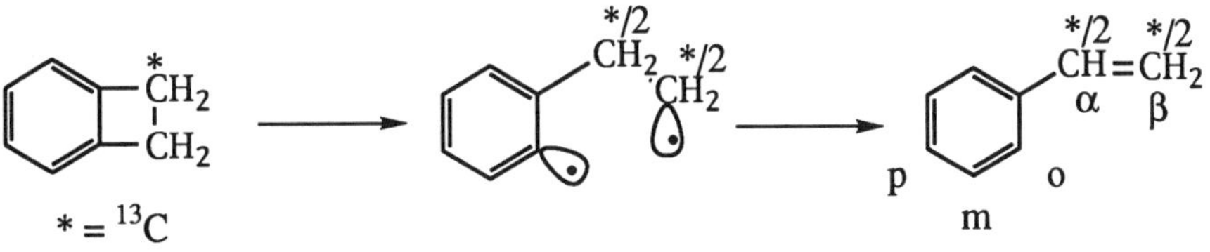

* = ^{13}C

Mechanism 2

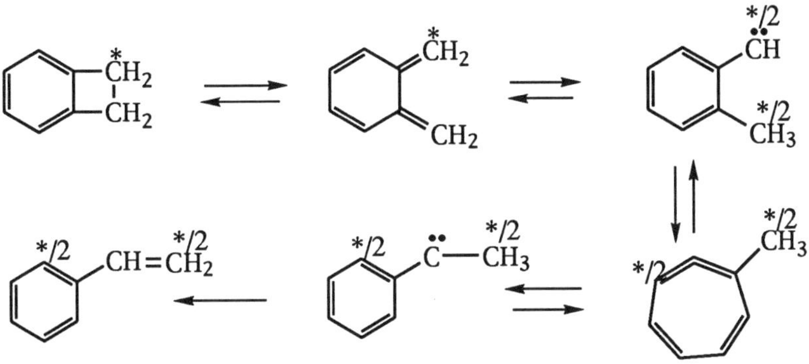

Scheme 2

The labeling results showed *beta* (48 %), *ortho* (30 %), *alpha* (14 %), *meta* (4 %), and *para* (4 %). These results are clearly consistent with our mechanism (Mechanism 2) as the dominant mechanism, but some minor processes must also intervene. The *alpha* label could come from the diradical mechanism, but neither mechanism that we considered (Scheme 2) accounts for the minor amounts of *meta* and *para* labels. As a control, we checked the possibility that styrene itself can rearrange under the conditions. In fact, to our great surprise we found that *beta*-labeled styrene gave *alpha*-labeled styrene in 4% yield (5).

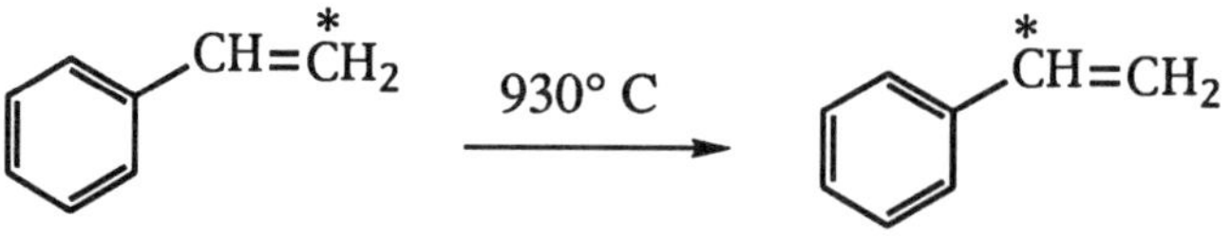

118 BENIGN BY DESIGN

$$\text{Ph-CH=}\overset{*}{\text{CH}}_2 \xrightarrow{930°\text{C}} \text{Ph-}\overset{*}{\text{CH}}\text{=CH}_2$$

This process accounts for at least some of the observed *alpha* label. Interconversion of *para*-tolylmethylene and *para*-xylylene and *meta*-xylylene could, however, introduce label into the *meta* and *para* positions of the styrene. These interconversions would occur in a loop that might exist in our mechanism.

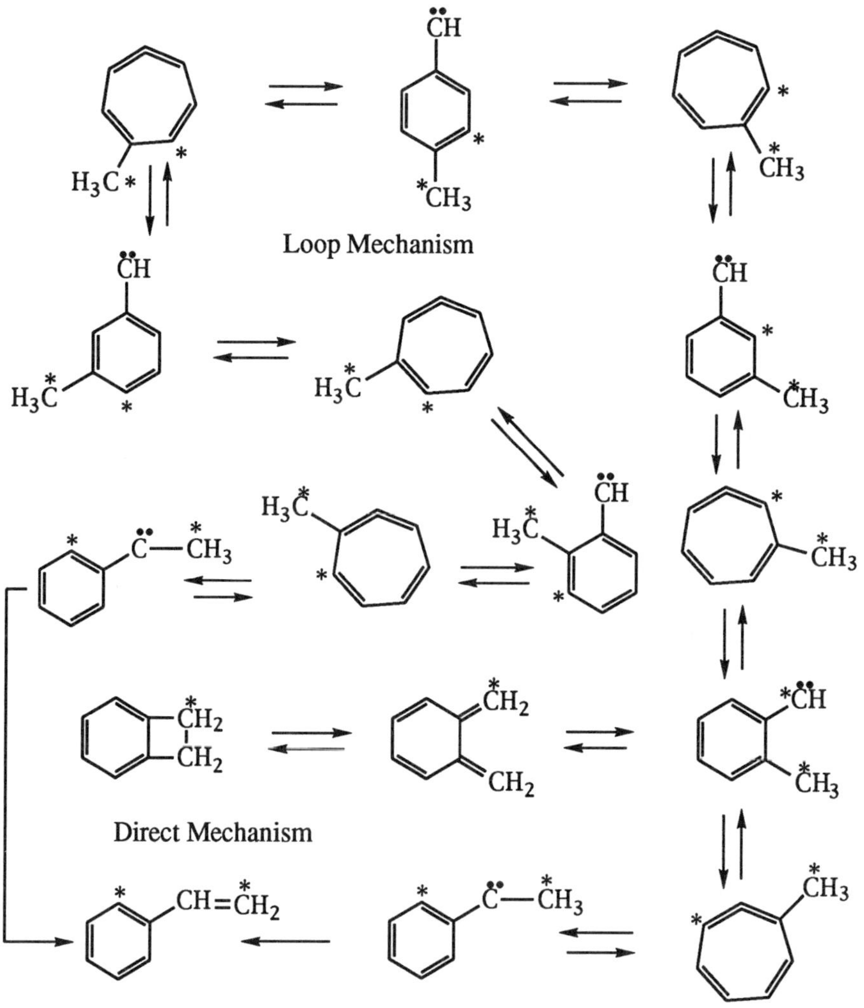

Loop Mechanism

Direct Mechanism

We could easily test for *para*-xylylene and *meta*-xylylene conversion to styrene, which would imply conversion of the xylylenes to the corresponding tolylmethylenes. Pyrolysis of [2.2]*para*-cyclophane gives *para*-xylylene and by inference [2.2]*meta*-cyclophane pyrolysis should give *meta*-xylylene. At 930° C, both cyclophanes ought to give styrene if the xylylenes convert to the tolylmethylenes.

In fact, high-temperature pyrolysis of both cyclophanes does give styrene (*5*). The principle of microscopic reversibility suggests that the xylylenes interconvert with the tolylmethylenes. Thermodynamics favor the xylylenes over the tolylmethylenes, and *ortho*-tolylmethylene certainly goes to *ortho*-xylylene (*10*). The loop mechanism with the interconversion of the xylylenes and the tolylmethylenes explains the minor labels in the *meta* and *para* positions of styrene (*5*).

The idea that *para*-tolylmethylene and *para*-xylylene interconvert suggested that we might use *para*-xylene as a precursor to styrene because Union Carbide Corporation has a patent that describes the thermal conversion of *para*-xylene to a high-temperature polymer presumably via *para*-xylylene. In fact, thermolysis of *para*-xylylene at high temperature does give styrene (*11*). In addition, *meta* and *ortho* xylenes also give styrene (*11*). Commercial, equilibrium-mixed xylenes, which contain some ethylbenzene, also give styrene (*11*). The equilibrium mixed xylenes are the cheapest aromatic available and are environmentally much safer than benzene.

An alternative to benzene and toluene as raw materials for styrene manufacture now exists. The process involves a single-step conversion that does not use a catalyst and gives 40% per pass yields of styrene. It uses a single, cheaper starting material that is environmentally safer than benzene. Physiological oxidation of the isomeric xylenes involves the methyl groups and does not make the arene oxide, which is the source of problems. The UCLA styrene process can probably use existing styrene plants, but a considerable development looms before this is a true commercial process. In fairness, the current styrene process development has invested roughly 10^5 times the dollars that we have spent in developing the UCLA styrene process. This study presents an interesting case of the United States research strategy in action. The National Science Foundation funded basic research directed toward solving an intriguing mechanistic problem, and the Environmental Protection Agency seeing promise in the results funded the development of the equilibrium mixed xylenes as a safer alternative to benzene. The fate of the UCLA styrene process rests on industry interest and on the political decision to take benzene out the global environment. If the decision is made to eliminate benzene, the UCLA styrene process offers an alternative. The simple fact that an alternative exists makes the political decision easier to make. Removing 13 billion pounds of benzene from global commerce can make a difference.

Acknowledgment

This work was supported by the National Science Foundation and by the Environmental Protection Agency.

Literature Cited

1. Innes, R. A.; Swift, H. E. *CHEMTECH* **1981**, 244-248.
2. Cava, M. P.; Deana, A. A. *J. Am. Chem. Soc.* **1959**, *81*, 4266-4268.
3. Baron, W. J.; DeCamp, M. R. *Tetrahedron Lett.* **1973**, 4225-4228.
4. Chapman, O. L.; Tsou, U.-P. E. *J. Am. Chem. Soc.* **1984**, *106*, 7974-7976.
5. Chapman, O. L.; Tsou, U.-P. E.; Johnson, J. W. *J. Am. Chem. Soc.* **1987**, *109*, 553-559.
6. West. P. R.; Chapman, O. L.; LeRoux, J.-P. *J. Am. Chem. Soc.* **1982**, *104*, 1779-1782.
7. McMahon, R. J.; Abelt, C. J.; Chapman, O. L.; Johnson, J. W.; Kreil, C. L.; LeRoux, J.-P.; Mooring, A. M.; West, P. R. *J. Am. Chem. Soc.* **1987**, 2456-2469.
8. Chapman, O. L.; McMahon, R. J.; West, P. R. *J. Am. Chem. Soc.* **1984**, *106*, 7973-7974.
9. Chapman, O. L.; Johnson, J. W.; McMahon, R. J.; West, P. R. *J. Am. Chem. Soc.* **1988**, *110*, 501-509.
10. McMahon, R. J.; Chapman, O. L. *J. Am. Chem. Soc.* **1987**, *109*, 683-692.
11. Chapman, O. L.; Tsou, U.-P. E.; U.S. Patent 4,554,782, October 1, 1985.

RECEIVED September 13, 1994

BENIGN CHEMISTRY:
INDUSTRIAL APPLICATIONS

Chapter 10

Generation of Urethanes and Isocyanates from Amines and Carbon Dioxide

Dennis Riley, William D. McGhee, and Thomas Waldman

Monsanto Company, 800 North Lindbergh Boulevard, St. Louis, MO 63167

Use of carbon dioxide as a reagent in place of phosgene offers the potential for an inexpensive, safer, and more selective route to polyurethanes. Traditionally, isocyanates, which are used for the production of polyurethanes, are generated from the reaction of a primary amine with the highly toxic agent phosgene. The reaction of CO_2 with either primary or secondary amines generates the carbamate anion. By understanding those factors which control the nucleophilicity of the carbamate anion, we have been able to generate urethanes quantitatively even with secondary amine-derived carbamate anions. We have also found that isocyanates can be generated quantitatively by the "dehydration" of carbamate anions derived from primary amines and carbon dioxide using a variety of dehydrating reagents. These processes are accomplished under very mild conditions; e. g., 1 atm CO_2 at 50-80°C. The factors which favor production of urethane or carbamate are discussed and examples of the utility of these new chemistries are presented.

The synthesis of urethanes and isocyanates using carbon dioxide as a key reagent in place of phosgene offers the potential for an inexpensive and safer route to these useful materials. Traditionally, isocyanates, which are the primary feedstocks for the production of polyurethanes (from the reaction of polyols with polyfunctional isocyanates), are generated via a number of routes including oxidative carbonylation of the corresponding amine (*1*), reductive carbonylation of a nitroaromatic amine (*2*), thermal cracking of a urethane to alcohol and isocyanate(*3,4*), and phosgenation of the corresponding amine (*5*). Of these, only phosgenation is practiced on a significant industrial scale (equations 1 and 2). This route is not without significant drawbacks, however, not the least of which are the highly toxic and corrosive

$$RNH_2 + COCl_2 \longrightarrow RNCO + 2\ HCl \quad (1)$$
$$\text{phosgene} \qquad \text{isocyanate}$$

$$RNCO + R'OH \longrightarrow RNHCO_2R' \quad (2)$$
$$\qquad\qquad\qquad\qquad \text{urethane}$$

nature of phosgene itself and the forcing conditions required for generation of the isocyanate (typically >180°C). As a result of the reaction conditions, only limited functionality can be incorporated into the amine precursor prior to generation of the isocyanate via phosgenation. This necessarily limits the type of amine feedstocks which can be utilized for urethane production. A further limitation of the phosgene route is the potential for chloride containing impurities, such as carbamoyl chlorides, due to incomplete conversion to the isocyanate.

The purpose of this paper is to describe some of our work utilizing carbon dioxide as a building block for the synthesis of organic molecules and, in particular, the synthesis of urethanes and isocyanates. Clearly, from an environmental, safety and possibly economic standpoint, the use of CO_2 would offer many attractive features as a reagent. As one can note in equation 2 shown above, the generic formula for a urethane contains the elements of CO_2 in the proper oxidation state. Thus, at first glance, it would seem that use of CO_2 could be a potential pathway for synthesis of urethanes, replacing phosgene, if methods could be developed for activating CO_2, a notoriously stable molecule.

It has long been known that CO_2 will react directly with primary or secondary amines generating their respective carbamate salts (6-8) (equation 3):

$$2\ RNH_2 + CO_2 \longrightarrow RNH_3^+\ {}^-O_2CNHR \quad (3)$$

The activation of CO_2 via this carbamate salt-forming reaction is very efficient, but the reactivity of these salts is largely dominated by electrophile attack on the nitrogen center of the carbamate, not on the oxygen. This means the attack by alkyl halides generates secondary or tertiary amines (9-13) and attack by acylating agents generates amides (14).

It is the control of the reactivity of this carbamate anion which has allowed us to generate in quantitative yields either urethanes or isocyanates directly. By understanding those factors which promote the nucleophilicity of the carbamate oxygen, we have discovered how to generate urethanes directly from either primary or secondary amines (Activated Carbon Dioxide Chemistry-I (ACDC-I) (12, 13, 15-18). A key aspect of this chemistry was our discovery that carbamate anions can serve as excellent nucleophiles, when a suitable tertiary amine base is utilized to generate the carbamate anion in the reaction of carbon dioxide with a primary or secondary amine (equation 4). By proper choice of the tertiary amine co-base, which in general does not react with CO_2 under the mild conditions employed, electrophilic attack on the oxygen can be directed exclusively to the oxygen of the

$$RNH_2 + CO_2 + R'_3N \longrightarrow R'_3NH^+ \; {}^-O_2CNHR \qquad (4)$$

carbamate salt (equation. 5):

$$R'_3NH^+ \; {}^-O_2CNHR + EX \longrightarrow RNHCO_2E + R'_3NH^+ X^- \qquad (5)$$

where EX = an electrophile such as an alkyl chloride

We have also found that isocyanates can be generated quantitatively by the apparent dehydration of carbamate anions derived from primary amines and CO_2 (ACDC-II) (*18, 19*). Both ACDC-I & II processes are accomplished rapidly under very mild conditions; e. g., 0-80 °C and 1 - 10 atm CO_2. During the course of our studies on the O-alkylation of carbamate anions (from primary or secondary amines and carbon dioxide, ACDC-I) to form urethanes we found that carbamate anions derived from primary amines react rapidly with electrophilic dehydrating agents to give the corresponding isocyanate in excellent yield (an example with an acid chloride is shown in equation 6). Essential to the successful generation of the isocyanate is the prevention of the formation of amides and the corresponding

$$RNHCO_2^- \; {}^+HNR_3 + R'COCl + R_3N \longrightarrow RNCO + 2\,R_3NH^+ + Cl^- + R'CO_2^- \qquad (6)$$

symmetric urea. A variety of "dehydrating" agents and organic bases effect this transformation and the mild reaction conditions tolerate a wide variety of functional groups on the amine precursor.

As noted above, the reaction of amines, carbon dioxide and electrophiles has been reported in the literature, but the products generally are derived from attack on the nitrogen center, liberating CO_2 and producing the substituted amine or amide. Both the ACDC-I and -II chemistries rely on the production of the carbamate anion from the reaction of carbon dioxide with the substrate amine. Very few reports have appeared which have addressed the issue of utilizing the carbamate anion as an oxygen- nucleophile in direct substitution reactions (*9-14*).

Results

Generation of Urethanes. We have developed chemistry which utilizes the oxygen center of the carbamate anion to construct products in excellent yields with virtually 100% selectivity under mild conditions from amines, CO_2 and alkyl chlorides using conditions which enhance the nucleophilic nature of the oxygen center of the carbamate anion. The key to this chemistry lies in the use of sterically hindered, powerful organic bases (e. g., penta-alkylguanidines) (20-25). Kinetic and mechanistic studies are presented which support the hypothesis that the carbamate anion will function effectively as an oxygen nucleophile when there is sufficient charge separation between the carbamate anion and its positive counterion. Sterically hindered guanidine bases satisfy this requirement and promote exclusive O-alkylation with a variety of alkyl chloride electrophiles. Additionally, reactivity studies with a variety of primary and secondary amines derived carbamates are also discussed. Surprisingly, primary amines have been found to form bis-CO_2 adducts (carbamate dianions: $RN(CO_2^-)_2$); consequently, the fastest rates to the urethane product ($RHNCO_2R$") are often achieved at low CO_2 pressure; i. e., < 5 atm. Examples of the application of this chemistry are presented with an emphasis on the generation of potentially useful materials. Additionally, we have focused on the ability of the chemistry to generate novel materials, both because secondary amines can be used to generate urethanes and because the conditions are so mild that many types of functionality in the substrate can be tolerated, which would not be tolerated via phosgenation (180°C, HCl evolution, etc.).

In Table I are shown the results for a series of urethane forming reactions using butyl amine as the substrate (1 M) and benzyl chloride (1 eqv.) as the electrophile (equation 7) in acetonitrile (75 ml) solvent at 40°C under 50 psig CO_2 pressure (unoptimized).

$$n\text{-BuNH}_2 + \text{Base} + CO_2 \xrightarrow{\text{PhCH}_2\text{Cl}} n\text{-BuNHCO}_2\text{CH}_2\text{Ph} + n\text{-BuNHCH}_2\text{Ph} \quad (7)$$

Clearly, the hindered tertiary amine guanidine base structure affords the desired carbamate in a highly selective manner, while other hindered tertiary amine bases are much less effective.

For this chemistry CO_2 is added to the acetonitrile solution containing the amine substrate and the guanidine base. This causes an immediate exothermic reaction generating the carbamate anion in solution. Rapid addition of benzyl chloride followed by heating the solution to 40°C gives the desired product urethane in >95% selectivity, and in the case of diethyl amine the reaction is complete in ~ 1/ hr.

In order to probe the effect of the substrate amine structure on this chemistry forming carbamates, a series of reactions were run under identical conditions to those described for Table I using benzyl chloride as the electrophile and tetramethylcyclohexyl guanidine as the co-base. The reaction rates are tabulated as relative rates with the rate for n-octyl amine set as 1. The results are summarized in Table II.

Table I. Base Effect on the Generation of Urethanes

Amine	% Yield* Urethane	% Yield* Benzylbutyl Amine
naphthalene-1,8-diamine (−NH₂, −NH₂)	0	>95
2,2,6,6-tetramethylpiperidine	7	>90
bicyclic amidine (N,N)	69	30
cyclohexyl-N=C(NEt₂)₂	92	7
cyclohexyl-N=C(NMe₂)₂	94	6

*GC Yield

Conductance studies in acetonitrile were performed using the triethyl ammonium and tetramethyl guanidinium salts of n-butyl carbamate. They showed that at the same concentrations (studied from 0.0075 M to 0.75 M) the guanidinium salt has a much greater conductance (>6:1) than the triethyl ammonium salt. This is indicative of greater ion pairing with the trialkyl ammonium salt than with the positive charge delocalized guanidinium salt. This suggests that at least some of the high selectivity for urethane formation (O-alkylation) observed with the sterically hindered guanidine bases is due to a less associated carbamate anionic oxygen.

The effect of CO_2 pressure was also studied under the same conditions using butyl carbamate in acetonitrile with benzyl chloride as the alkylating agent (equation 7). The results are shown in Figure 1 below and demonstrate the unexpected result that increasing CO_2 pressure decreases the rate of the reaction.

Table II. Relative Rate of Reaction of Various Carbamate Anions

Amine	Relative Rate
n-octyl amine	1
n-butyl amine	1
cyclohexylmethyl amine	1.5
aniline	8
s-butyl amine	12
cyclohexyl amine	14.5
t-butyl amine	16
di-n-butyl amine	29
diethyl amine	32
acetate	60

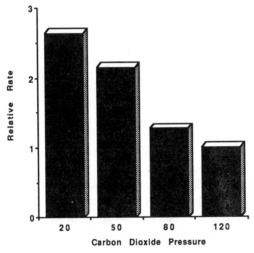

Figure 1. Effect of CO_2 Pressure on Rate of Reaction

A possible explanation of this observation is that a second molecule of carbon dioxide adds to the carbamate forming a double insertion product (dianion) which is less reactive than the carbamate mono anion. In order to test this we probed the addition of $^{13}CO_2$ to both $Et^{15}NH_2$ and $Et_2^{15}NH$ by ^{15}N NMR spectroscopy. The appearance of a doublet at δ 87.6 ppm for the addition of $^{13}CO_2$ to $Et_2^{15}NH$ with cyclohexyltetramethylguanidine (CyTMG) as the co-base indicates the generation of the monocarbamate salt, $Et_2{}^{15}N^{13}CO_2{}^-$ ^+HCyTMG with J_{N-C} = 19.3 Hz. The appearance of a triplet centered at δ 129.1 in the ^{15}N NMR of the product of the addition of $^{13}CO_2$ to $Et^{15}NH_2$ with two equivalents of CyTMG indicates complete formation of a bis CO_2 addition product: $Et^{15}NC(^{13}CO_2{}^-)_2$ $(^+HCyTMG)_2$ with J_{N-C}=15.2 ppm. These results point out a very important fact that secondary amines form only the mono addition product, but the less sterically hindered primary amines form the bis CO_2 adduct. Clearly, CO_2 pressure is as an important variable in the optimization of the reaction.

We have explored the scope of this reaction with many different types of amines, including aromatic amines, di- and tri-functional aliphatic amines and chiral amino acids. In all examples very high yields of the corresponding urethanes can be made using a variety of alkyl chlorides as the electrophile, including, ethyl chloride, benzyl chloride, etc.. The reaction with amino acids to prepare the N-carbobenzoxy carbonyl protecting group proceeds with complete retention of configuration.

Generation of Isocyanates. We have found that the carbamate anions described above will undergo a rapid reaction with classical dehydration reagents to yield isocyanates. For example, generation of carbamate anions is readily accomplished by the addition of CO_2 (1 - 5 atm) to a solution of the primary amine and at least 2 equivalents of a tertiary amine organic base, e. g., NEt_3. Unlike the urethane forming reaction, the co-base does not need to be a guanidine to achieve high yields of isocyanate. In general triethyl ammonium carbamates form slurries, regardless of solvent, whereas the pentaalkyl guanidinium salts give soluble carbamate salts. In either case, the addition of a stoichiometric amount of a dehydration agent, such as PCl_3, $SOCl_2$, $POCl_3$, etc., per amine group to the reaction mixture gives an exothermic reaction from which the desired isocyanate can be isolated in excellent yield (typically >90%). In general the reaction, to afford isocyanate, occurs virtually instantaneously at 0°C with such a dehydration agent. Many dehydration agents work well; including non-halide containing agents such as P_4O_{10}, SO_3, and various organic acid anhydrides. Examples of the utility of this technology are shown in Table III for the reaction shown in equation 8:

$$n\text{-}C_8H_{17}NH_2 + 2 \text{ Base} + CO_2 \xrightarrow[\text{0°C, 1 eqv. dehydration agent}]{\text{acetonitrile}} n\text{-}C_8H_{17}NCO \quad (8)$$

Conditions: $n\text{-}C_8H_{17}NH_2$ (5 mmol), base (10 mmol), 25 ml solvent 1hr reaction time at 25°C.

Table III: Production of Octyl Isocyanate Via Dehydration of Carbamates

Base	CO_2 Pressure (psig)	Dehydration Agent	% Yield Isocyanate (% Urea)
NEt_3	15	$POCl_3$	98 (--)
NEt_3	15	PCl_3	96 (--)
NEt_3	15	SO_3	99 (--)
CyTMG[a]	80	$(CF_3CO)_2O$	91 (7)
CyTMG[a]	15	P_4O_{10}	75 (22)
NEt_3	15	$P_4O_{10}(2\ NEt_3)$[b]	97 (trace)
CyTMG[a] (CyNCS)	75	$SOCl_2$	70 (30%
CyTMG	15	$(CH_3CO)_2O$	18[c]

[a]CH_2Cl_2 solvent. [b]Dissolution/activation of P_4O_{10} with NEt_3 (26). [c]By-product is the n-octylacetamide.

The last entry in Table III merits some additional attention. As in the case of the direct alkylation of carbamate anions, the chemistry of these apparent dehydration reactions proceeds through formation of oxygen attack on the carbamate by the electrophilic dehydration agent. For example, with acid anhydrides the attack of the electrophile on the carbamate anion can proceed either through attack on the nitrogen center to generate amide products or on the oxygen center to generate a mixed anhydride, which in the presence of an additional mole of tertiary amine base yields isocyanate and the ammonium salt of the organic acid (27). Not surprisingly, the selectivity for formation of isocyanate products is highly dependent on the nature of the organic base as well as the acid anhydride. This mechanism is depicted in the Figure 2 shown below. The major pathway for formation of the amide by-product appears to be the direct attack of the acid anhydride on the free amine, as monitored by NMR. Since high CO_2 favors the unreactive bis-CO_2 adduct, high selectivities to the desired isocyanate products can be very sensitive to an optimization of the reaction conditions.

The extension of this methodology to the synthesis of polyisocyanates of commercial interest has been done. Importantly, due to the extremely mild conditions under which this chemistry can be conducted, there are many potential substrates which can be converted to their isocyanates which could never survive the conditions present in a phosgenation reaction. Examples of polyisocyanates and unique isocyanates are shown below in Figure 3: all examples shown here have been prepared by the general methods and conditions documented in the examples shown in Table III and they have been prepared in >90% selectivities.

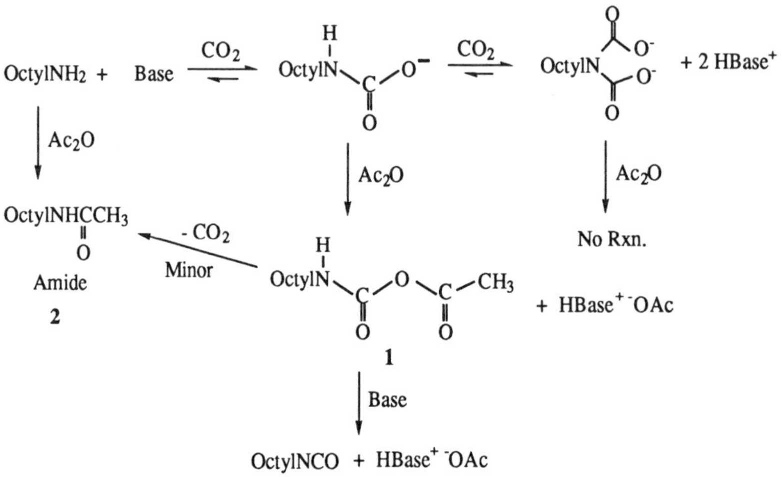

Figure 2: Proposed Pathway for Reaction of Octylamine Carbamate with Acetic Anhydride

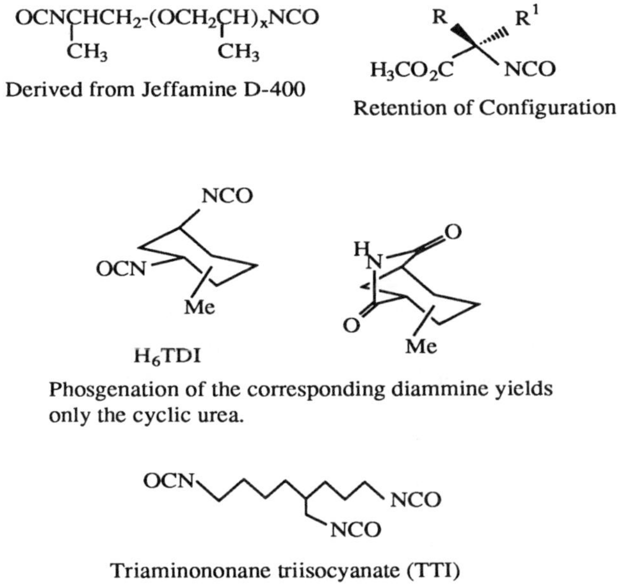

Derived from Jeffamine D-400: $OCNCHCH_2-(OCH_2CH)_xNCO$ with CH_3 groups

Retention of Configuration: R, R^1, H_3CO_2C, NCO

H_6TDI

Phosgenation of the corresponding diammine yields only the cyclic urea.

Triaminononane triisocyanate (TTI)

Figure 3. Novel Isocyanates Generated Via ACDC-II

Discussion and Summary

A very key aspect to the development of the chemistries discussed here is the recycle of the tertiary amine co-base which is utilized. Details such as which base, reaction solvent, the mode of recovery, etc. are all critical factors, but the optimization of conditions critically depends on the substrate and the type of chemistry one is attempting; i.e., ACDC-I or ACDC-II. With ACDC-II the technology permits the synthesis of a wide variety of isocyanates by a completely halide-free route using acid anhydrides as the electrophilic, oxophilic dehydration agent. Use of dehydration agents, which can be recycled back to their anhydride form from their acid form, permits, at least conceptually, the development of a completely waste-free route to isocyanates using electrolytic salt-splitting technology.

In summary, we have described in this presentation a new CO_2-based non-phosgene route to isocyanates and urethanes which allows us to synthesize molecules possessing these functionalities under very mild conditions in very high selectivity. The very mild nature of the reaction conditions makes it possible to synthesize isocyanates or urethanes from amines which normally would not survive the extreme conditions (high temperature, HCl evolution, etc.) of a phosgenation reaction. Consequently, highly functionalized urethanes or amines can be prepared. Further, urethanes can be readily prepared from secondary amines providing a facile route to secondary amine-based urethanes--materials which cannot be generated via phosgenation. High-purity isocyanates can also be made via a non-halide route, using dehydrating agents which are halide free, such as acid anhydrides. This offers the potential for developing a non-salt and non-halide waste generating route to isocyanates.

Literature Cited

1. Leung, T.W.; Dombek, B.D. *J. Chem. Soc., Chem. Commun.* **1992**, 205-206.
2. Valli, V.L.K.; Alper, H. *J. Am. Chem. Soc.* **1993**, *115*, 3778-3779.
3. Merger, F.; Towae, F. U.S. Patent # 4,713,476, 1987.
4. Sundermann, R.; Konig, K.; Engbert, T.; Becker, G.; Hammen, G. U.S. Patent # 4,388,246, 1983.
5. *Polyurethane Handbook*; Oertel, G., Ed.; Hanser Publishers; Munich, 1985.
6. Fichter, R.; Becker, B. *Chem. Ber.* **1911**, *44*, 3481-3485.
7. Jensen, A.; Christensen, R.; Faurholt, C. *Acta Chem. Scand.* **1952**, *6*, 1086-1089 and references therein.
8. Lallau, J.P.; Masson, J.; Guerin, H.; Roger, M.-F. *Bull. Soc. Chim. Fr.* **1972**, 3111-3112.
9. Bruneau, C.; Dixneuf, P.H. *J. Mol. Catal.* **1992**, *74*, 97-107.
10. Aresta, M.; Quaranta, E. *Tetrahedron,* **1992**, *48*, 1515-1530.
11. Aresta, M.; Quaranta, E. *J. Chem. Soc., Dalton Trans.* **1992**, 1893-1899.
12. McGhee, W.D.; Riley, D.P. *Organometallics* **1992**, *11*, 900-907.

13. McGhee, W.D.; Riley, D.P.; Christ, M.E.; Christ, K.M. *Organometallics* **1993**, *12*, 1429-1433 and references cited therein.
14. Belforte, A.; Dell'Amico, B.D.; Calderazzo, F. *Chem. Ber.* **1988**, *121*, 1891-1897.
15. Riley, D.P.; McGhee, W.D. U.S. Patent # 5,055,577, 1991.
16. Riley, D.P.; McGhee, W.D. U.S. Patent # 5,200,547, 1993.
17. McGhee, W.D.; Parnas, B.L.; Riley, D.P.; Talley, J.J. U.S. Patent # 5,223,638, 1993.
18. McGhee, W.D.; Stern, M.K.; Waldman, T.E. U.S. Patent # 5,233,010, 1993.
19. McGhee, W.D.; Waldman, T.E. U.S. Patent # 5,189,205, 1993.
20. Bredereck, H.; Bredereck, K. *Chem. Ber.* **1961**, *94*, 2278-2295.
21. Barton, D.H.R.; Elliott; Gero, S.D. *J. Chem. Soc., Perkin Trans. 1* **1982**, 2085-2090.
22. Oszczapowicz, J.; Raczynska, E. *J. Chem. Soc., Perkin Trans. 2* **1984**, 1643-1646.
23. Schwesinger, R. *Chimia* **1985**, *39*, 269-272.
24. Leffek, K.T.; Pruszynski, R.; Thanapaalasingham, K. *Can. J. Chem.* **1989**, *67*, 590-595.
25. Boyle, P.H.; Convery, M.A.; Davis, A.P.; Hosken, G.D.; Murray, B.A. *J. Chem. Soc., Chem. Commun.* **1992**, 239-242.
26. Cherbuliez, E.; Leber, J.P., Schwarz, M., *Helv. Chim. Acta* **1953**, *36*, 1189-1199.
27. Motoki, S.; Saito, T.; Kagami, H. *Bull. Chem. Soc. Jpn.* **1974**, *47*, 775-776 and references cited therein.

RECEIVED August 4, 1994

Chapter 11

Nucleophilic Aromatic Substitution for Hydrogen
New Halide-Free Routes for Production of Aromatic Amines

Michael K. Stern

Monsanto Company, 800 North Lindbergh Boulevard, St. Louis, MO 63167

The development of environmentally favorable routes for the production of commercially relevant chemical intermediates and products is an area of considerable interest to the chemical processing industry. These synthetic routes will ideally focus on elimination of waste at the source and will require, in most cases, the discovery of new atomically efficient chemical reactions. In light of these emerging requirements, we have focused our attention on nucleophilic aromatic substitution for hydrogen (NASH) reactions as a means to generate aromatic amines without the need for halogenated materials or intermediates. Two new routes based on NASH reactions are described for the manufacturing of 4-aminodiphenylamine (4-ADPA) and 4-nitroaniline (PNA) which illustrate the concept of alternate chemical design. In addition, mechanistic studies into the oxidation of the critical σ-complex intermediates are presented.

One of the oldest practiced industrial chemical reactions is the activation of benzene by chlorination (Figure 1). The resulting chlorobenzenes can be further activated towards nucleophilic aromatic substitution by nitration producing a mixture of *ortho* and *para*-nitrochlorobenzene (PNCB), **1**. These intermediates are employed in a variety of commercial processes for production of substituted aromatic amines. For instance, in the simplest case when PNCB is reacted with ammonia, the corresponding nitroanilines are produced. Since neither chlorine atom ultimately resides in the final product, the ratio of pounds of by-products produced per pound of product generated is highly unfavorable. In addition, these processes typically generate aqueous waste streams which contain high levels of inorganic salts that are difficult and expensive to treat.

By contrast, a more direct and atomically efficient route for the production of aromatic amines would be to eliminate the need for halogenation of benzene. This can be achieved by a class of reaction illustrated in Figure 2 which is known as

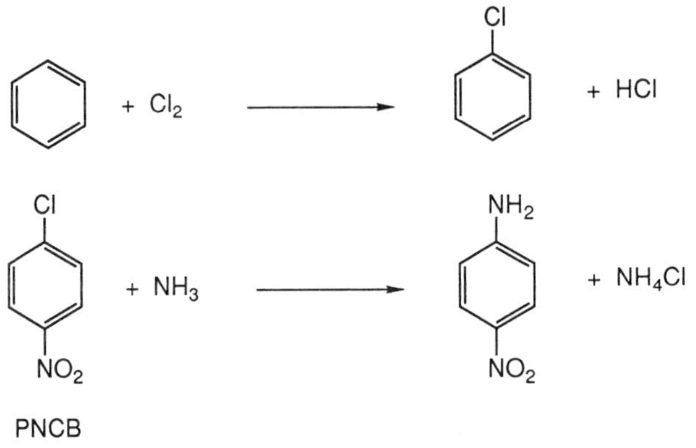

Figure 1. Production of aromatic amines from chlorobenzene.

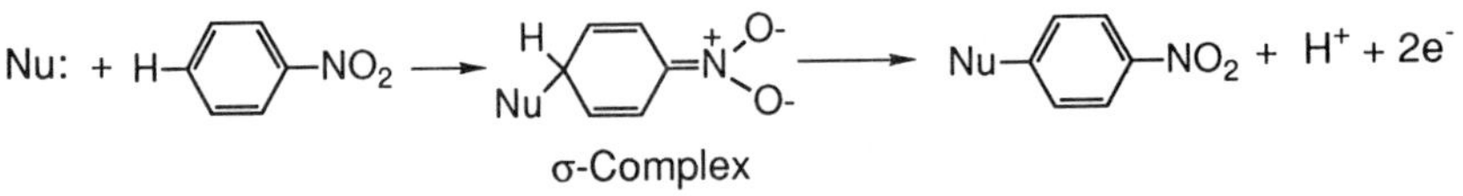

Figure 2. Nucleophilic aromatic substitution for hydrogen reaction.

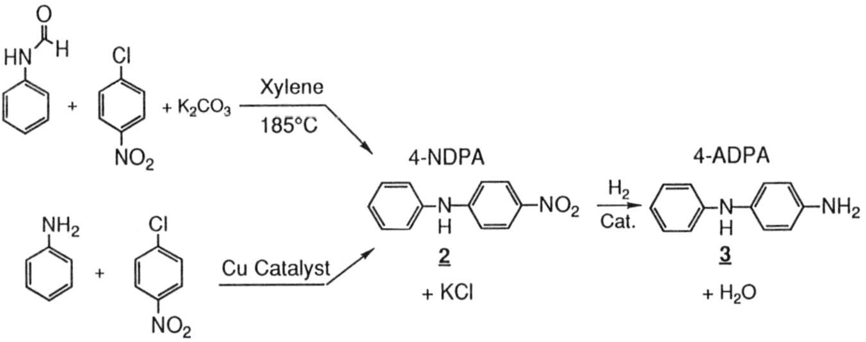

Figure 3. Currently practiced commercial routes to 4-aminodiphenylamine.

nucleophilic aromatic substitution for hydrogen (NASH). While these types of reaction have been known for nearly 100 years (1), this chemistry generally proceeds in low yields, gives mixtures of *ortho* and *para* substitution products, and requires the use of environmentally unfavorable external oxidants (2). This chapter focuses on two new examples of NASH chemistry which are applicable to the production of commercially relevant aromatic amines.

A New Route to 4-Aminodiphenylamine (4-ADPA)

4-Aminodiphenylamine is a key intermediate in the manufacturing of antioxidants used as additives to rubber products particularly tires. Current processes for the manufacturing of 4-ADPA are heavily dependent on halogenated reagents which ultimately generate waste streams laden with inorganic salts and trace amounts of organic by-products. The reaction of aniline or aniline derivatives with PNCB shown in Figure 3 is the critical coupling reaction currently practiced in the manufacturing of 4-nitrodiphenylamine (4-NDPA), **2**. Hydrogenation of **2** produces 4-ADPA, **3**, which is a key intermediate in the *para*-phenylenediamine class of antioxidant used in rubber products. This process suffers all the problems outlined above since it relies on chlorine to activate the aromatic ring towards nucleophilic attack. It is estimated that for every pound of 4-ADPA produced via this route 1.0-1.5 pounds of waste are generated. It has recently been shown that the base catalyzed NASH reaction of aniline and nitrobenzene can be used to generate intermediates like **2** and, therefore, offers an alternate route for the production of 4-ADPA (3).

We observed that the addition of nitrobenzene to an aniline solution containing 2.2 equiv. of tetramethylammonium hydroxide dihydrate (TMA(H)·2H2O) at 50°C under anaerobic conditions caused the immediate formation of a red species. Analysis of the reaction mixture indicated that 4-nitrosodiphenylamine (4-NODPA), **4**, and 4-NDPA, **2**, were generated in 89% and 4% yield, respectively. In addition, small amounts of azobenzene, **5**, (3.5%), and phenazine, **6**, (3.5%) were produced. Spectrophotometric analysis of a reaction mixture containing equal molar amounts of aniline, nitrobenzene and TMA(H)·2H2O in DMSO revealed a single broad absorbance with a λ_{MAX}=494 nm which is indicative of the deprotonated form of the nitroaromatic amines (4). Thus, it was concluded that the primary products of this reaction are not 4-NODPA, **4**, or 4-NDPA, **2**, but rather their tetramethylammonium salts, **7**, and, **8**, respectively.

Our investigations revealed that the amount of water present, the ratio of aniline to nitrobenzene, and whether the reaction was conducted in the presence or absence of O_2 had a critical effect on selectivity and overall yield of this reaction. The effect that varying the aniline:nitrobenzene ratio (AN/NB) has on product distributions is summarized in Table I. These results clearly demonstrate that as AN/NB is increased, the yield of **7** increases at the expense of **8**.

That water has a profound effect on product yield and selectivity was demonstrated by a series of reactions that contained varying amounts of protic material. Thus, as the amount of water was increased in the reaction, the conversion of nitrobenzene decreased dramatically. However, increasing the amount of water also resulted in reactions which displayed a higher degree of selectivity to the desired *para*-substituted products (Figure 4). Accordingly, it was clear that the proper control of water would be critical if high yields and selectivity of the desired products were to be achieved.

The effect that O_2 has on the reaction was illustrated by a series of reactions run at various temperatures under aerobic conditions. The results of these studies are summarized in Table II.

Table I: Effect of Aniline/Nitrobenzene Mole Ratio on Product Distribution

Mole Ratio Aniline/Nitrobenzene	% of Total	
	<u>7</u>	<u>8</u>
1.32	15	80
11.9	55	35
51.5	86	9

Table II: Effect of Temperature on Yield and Selectivity Under Aerobic Conditions

Temp. °C	% Conversion Nitrobenzene	4-NODPA	% Yields[a] 4-NDPA	Azobenzene	Phenazine
23	73	51	8.5	12	trace
50	98	86	7.6	21	1.6
80	100	89	7.0	55	2.0

[a]Yields based on Moles nitrobenzene charged.

The conversion of nitrobenzene to 4-NODPA and 4-NDPA increased with temperature. However, the molar yield of azobezene also increased with rising temperature indicating that under aerobic conditions there must be an oxgyen dependant route for the production of azobenzene that does not consume nitrobenzene. That azobenzene was formed by two separate pathways in the presence and absence of O_2 was confirmed by a series of deuterium labeling experiments. When aniline-d_5 was used as solvent and the reaction was conducted in the absence of O_2, 4-NODPA-d_5, 4-NDPA-d_5 and azobenzene-d_5 were observed. However, when the identical reaction was conducted in the presence of O_2 a significant amount of azobenzene-d_{10} was produced indicating that under aerobic conditions the predominant route for the formation of azobezene is via the oxidative coupling of aniline (5).

A mechanism consistent with these experimental observations is shown in Figure 5. Deprotonation of aniline generates the anilide ion which is required for

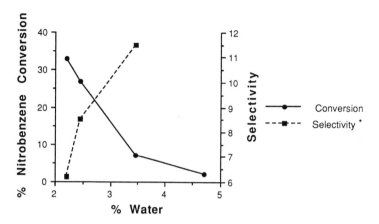

Figure 4. Effect of water on nitrobenzene conversion and reaction selectivity.
*Selectivity is defined as the ratio of *para* substituted products to *ortho* substituted products.

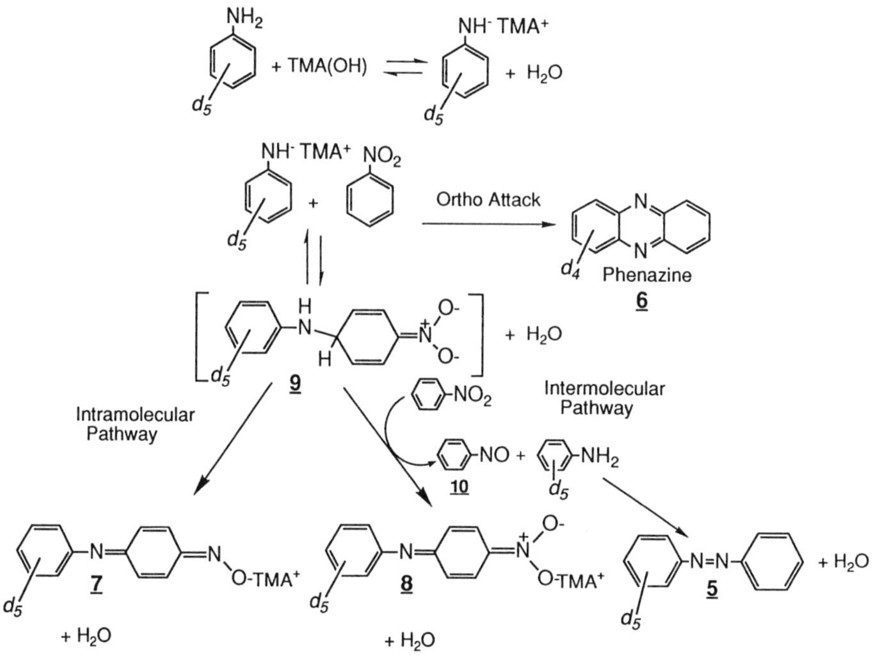

Figure 5. Proposed mechanism for the reaction of aniline and nitrobenzene.

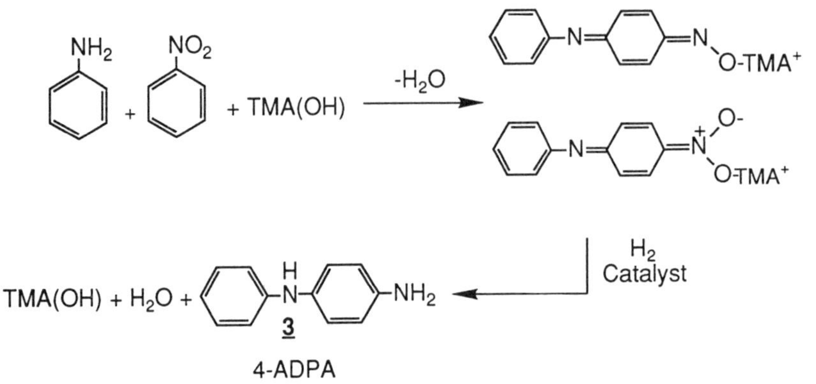

Figure 6. New route to 4-aminodiphenylamine based on NASH reaction.

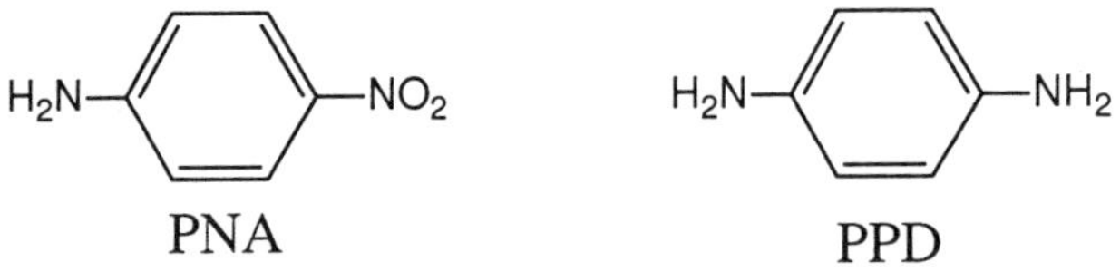

PNA

PPD

Figure 7. Commercially important aromatic amines.

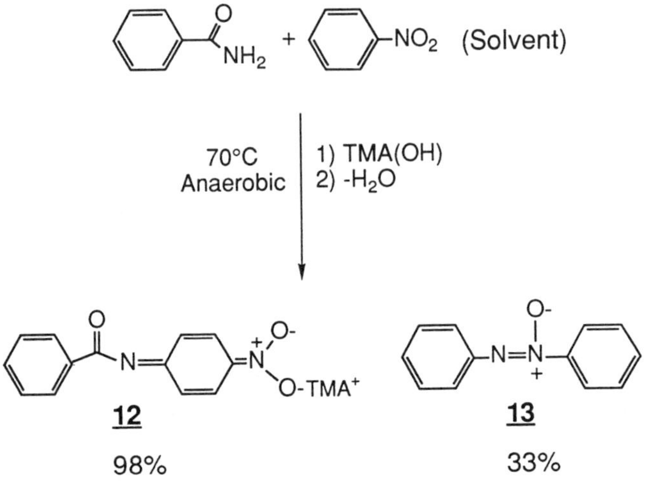

Figure 8. NASH reaction between benzamide and nitrobenzene.

nucleophilic attack on nitrobenzene forming the σ-complex **9** (6). We interpret the inhibitory effect of water to result from its effect on the equilibrium between aniline and anilide. Formal oxidation of **9** can occur by two separate pathways: 1) an intramolecular redox process with the nitro group of **9** functioning as the oxidizing agent generating **2** and **7**; 2) an intermolecular pathway with free nitrobenzene functioning as the oxidant producing nitrosobenzene, **8** and **10**. Nitrosobenzene is not directly observed since it will condense with aniline under the reaction conditions to produce azobenzene. In either pathway, a nitro group is formally reduced to a nitroso moiety by the hydride leaving group. The small amount of phenazine observed results from *ortho* attack on nitrobenzene followed by a similar intramolecular oxidation and ring closure. Furthermore, the observation that equal molar amounts of **8** and azobenzene **5** are produced is consistent with this mechanism.

Understanding the mechanism of the coupling reaction and, in particular, the oxidation of **9** was critical in moving this technology out of the laboratory and into a pilot plant facility. This reaction forms the basis for a new process which could be used to manufacture 4-ADPA (Figure 6). This route will generate approximately 0.07 pounds of waste per pound of 4-ADPA produced with the complete elimination of inorganic salt waste streams. The major by-product of this process is distilled water. These numbers can be put into perspective when compared to the waste generated by the current processes which typically falls in the range of 1 to 1.5 pounds per pound 4-APDA produced. Accordingly the new technology based on NASH reactions is an attractive and environmentally favorable alternative for the commercial production of 4-APDA.

An Alternate Route to *p*-Nitroaniline and *p*-Phenylenediamine

Another commercially important pair of aromatic amines are *p*-nitroaniline (PNA) and its derivative *p*-phenylenediamine (PPD) which are shown in Figure 7. PNA is currently produced at Monsanto by the reaction of ammonia with PNCB, as described above, and can be catalytically hydrogenated to PPD. Recently a new example of NASH chemistry directly applicable to the production of PNA and PPD was discovered (7).

We have shown that the reaction of benzamide, **11**, and excess nitrobenzene in the presence of base under anaerobic conditions generated the tetramethylammonium salt of N-(4-nitrophenyl)- benzamide, **12**, in 98% yield under mild conditions (7). The only other observable product in this reaction was azoxybenzene, **13**, which was formed in 33% yield based on benzamide charged (Figure 8). That azoxybenzene is observed as a by-product of this reaction under anaerobic conditions indicates that nitrobenzene is functioning as the primary oxidant in this system. This was confirmed by the analysis of the reaction mixture by EPR spectroscopy which revealed that nitrobenzene radical anion was also formed during the course of the reaction as evident by its distinct 54 line pattern (8).

A mechanism which explains the simultaneous formation of **12** and **13** is shown in Figure 9. Intermolecular oxidation of σ-complex **14** by nitrobenzene generates **12** and initially nitrobenzene radical anion, **15**, which can decompose by a variety of pathways including disproportionation to give nitrosobenzene (9). The

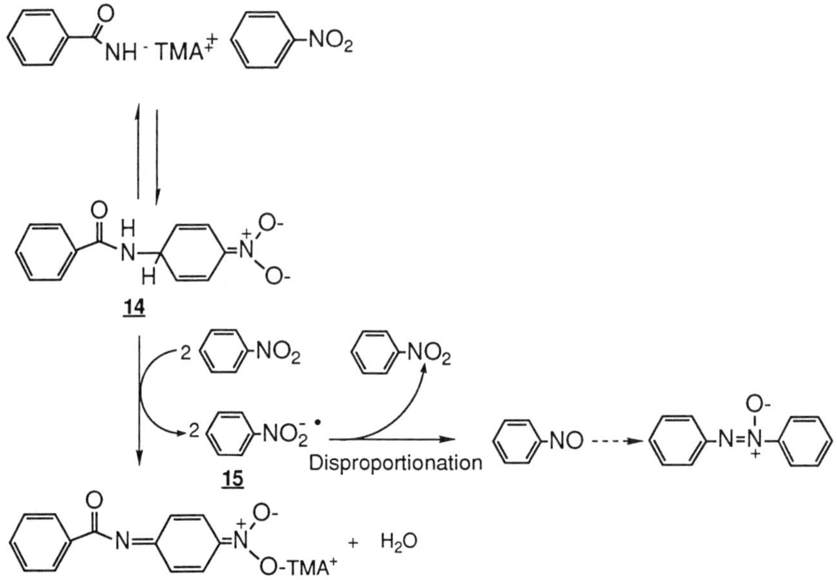

140 BENIGN BY DESIGN

Figure 9. Proposed mechanism for the reaction of benzamide and nitrobenzene under anaerobic conditions.

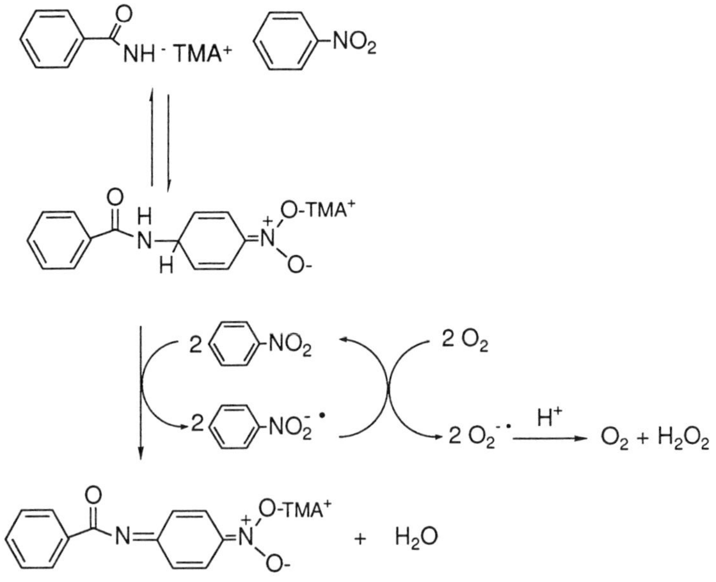

Figure 10. Proposed mechanism for the reaction of benzamide and nitrobenzene under aerobic conditions.

ultimate formation of azoxybenzene is then governed by a cascade of electron transfer and nucleophilic reactions between the radical anions of nitrobenzene, nitrosobenzene and N-hydroxyaniline, all of which can be directly accessed in the reaction mixture by the reduction of nitrobenzene to various extents (*10*). The overall stoichiometry required by this mechanism (equation 1) is consistent with the observed yields of **12** and **13**.

$$\text{PhC(O)NH}_2 + 5/3 \text{ PhNO}_2 + \text{TMA(OH)} \longrightarrow \mathbf{13} + 1/3\,\mathbf{12} + 2\,H_2O \quad (1)$$

In contrast to the case where aniline is used as the nucleophile, the benzamide reaction can be improved by utilizing dioxygen in the reaction mixture since **11** is resistant to autoxidation. Under aerobic conditions the nitrobenzene radical anion is readily trapped by O_2 generating superoxide and nitrobenzene (Figure 10) (*11*). This reaction pathway inhibits the formation of azoxybenzene by diverting the electron transfer cascade and ultimately utilizing dioxygen as the terminal oxidant. Thus, under aerobic reaction conditions **12** is the only observed reaction product.

The ability to prepare **12** in high yield and regioselectivity affords a novel route for the preparation of PNA and its derivative PPD (Figure 11). The addition of water to **12** results in protonation of the amide nitrogen producing **16** and regenerating TMA(OH). Subsequent reaction of **16** with ammonia cleaves the amide bond generating PNA which can be hydrogenated to PPD under catalytic conditions. It is important to note that this route would be catalytic in TMA(OH) and also allows for the recycling of the amide nucleophile. Accordingly, the overall stoichiometry for this series of reactions illustrates the formal amination of nitrobenzene with ammonia.

The formation of substituted anilides from the reaction of amides with nitrobenzene is the first example of the direct formation of aromatic amide bonds via nucleophilic aromatic substitution for hydrogen. This reaction proceeds in high yield and regioselectivity, and does not require the use of halogenated materials or auxiliary leaving groups. Furthermore, these studies have demonstrated that the use of O_2 as the terminal oxidant in NASH reactions can result in a highly selective and environmentally favorable route for the production of PNA and PPD.

Conclusion

Two examples of nucleophilic aromatic substitution for hydrogen reactions were described from which we have proposed two new atomically efficient processes for the manufacturing of commercially relevant aromatic amines. Our mechanistic studies have revealed that the direct oxidation of σ-complex intermediates by either nitro groups or O_2 can eliminate the need for chlorination of benzene as a starting point for the manufacturing of aromatic amines. Accordingly, these reactions demonstrate the key objective of alternate chemical design which is not to make the waste in the first place.

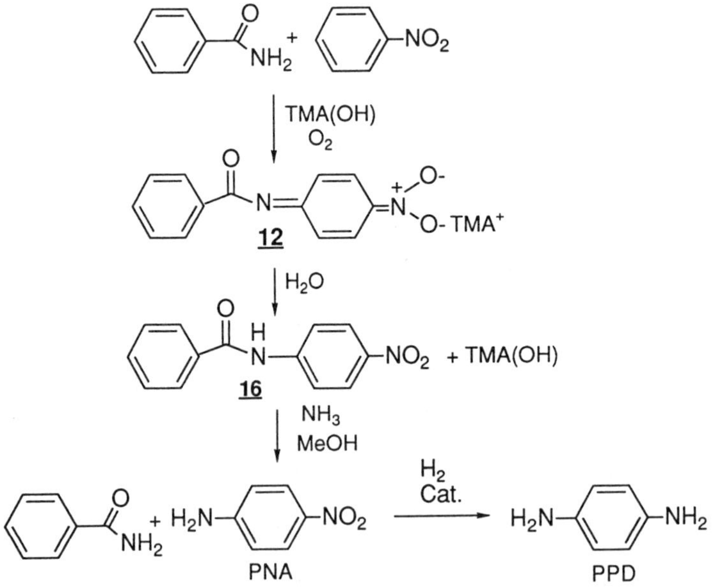

Figure 11. New route to PNA and PPD based on NASH reaction.

Acknowledgments

The author would like to thank Dr. Fred Hileman and Mr. Brian Cheng for their contributions to this work.

Literature Cited

1. Wohl, A. *Chem. Ber.* **1903**, *36*, 4235.
2. Terrier, F. In *Organic Nitro Chemistry Series*; Feuer, H., Ed.; VCH Publishers, Inc.: New York, NY, 1991 and references therein.
3. Stern, M. K.; Hileman, F. D.; Bashkin, J. K. *J. Am. Chem. Soc.* **1992**, *114*, 9237-9238.
4. Langford, C. H.; Burwell, R. L., Jr. *J. Am. Chem. Soc.* **1960**, *82*, 1503.
5. Jeon, S.; Sawyer, D.T. *Inorg. Chem.* **1990**, *29*, 4612-4615.
6. Buncel, E., In *Supplement F: The Chemistry of Amino, Nitroso, and Nitro Compounds and Their Derivatives*; Patai, S., Ed.; Wiley: London, **1982**; Part 2, Chapter 27.
7. Stern, M.K.; Cheng, B. K. *J. Org. Chem.* **1993**, *58*, 6883-6888.
8. Geske, D. H.; Maki, A. H. *J. Am. Chem. Soc.* **1960**, *82*, 2671-2676.
9. Russel, G. A.; Janzen, E. G.; Strom, E. T. *J. Am. Chem. Soc.* **1964**, *86*, 1807-1814.
10. Russel, G. A.; Janzen, E. G. *J. Am. Chem. Soc.* **1962**, *84*, 4153-4154.
11. Guthrie, R. D.; Nutter, D. E. *J. Am. Chem. Soc.* **1982**, *104*, 7478-7482.

RECEIVED August 4, 1994

Chapter 12

Chemistry and Catalysis
Keys to Environmentally Safer Processes

Leo E. Manzer

DuPont, Corporate Catalysis Center, Central Science and Engineering, Wilmington, DE 19880−0262

The public and industry are becoming much more concerned over environmental issues. The effect will clearly increase the relative importance of environmental costs vs feedstock costs for current and future plants. As a result, several issues will become increasingly important and require significant research effort: processes with 100% yield (by whatever process); catalyst recovery, regeneration and recycling; heterogenization of homogeneous catalysts; chiral pharmaceutical and agrichemicals; polymer recycling and environmentally safer processes. Several of these issues will be highlighted in this paper using examples from DuPont R&D.

The greatest impact of catalysis will be in the development of new processes with essentially zero waste. Clearly the best approach is to develop processes with very high single pass yields. Often this is not possible so all waste and byproducts must be handled in an environmentally acceptable fashion. Some waste can be incinerated. Another approach is to convert it to salable products. Catalysis plays a key role in the conversion to salable products as illustrated by examples from our nylon 6.6 process.

High Yield-Low Waste Processes

Elimination of byproducts and process waste is becoming a major issue and will clearly determine the viability of future chemical processes. Those processes which strive for zero emissions and very high process yields at the lowest cost will be winners in the 21st century. It is difficult to project the total effect of environmental issues much into the future, other than to say that they will become increasingly important. One example of a very high yield process involves the synthesis of HCFC-141b, equation 1, a product from DuPont's research and development on alternatives to the ozone depleting chlorofluorocarbons (CFCs). It is a replacement for several solvent blends and azeotropes, and for blowing foams:

$$CH=CCl_2 + HF \longrightarrow CH_3CFCl_2 + CH_3CF_2Cl + CH_3CF_3 \quad (1)$$
$$VCl_2 \qquad\qquad\qquad 141b \qquad 142b \qquad 143a$$

Historically, low yields to 141b have been reported since the reaction is thermodynamically in favor of HFC-143a. We discovered (*1*) that a specially prepared AlF_3 catalyst was remarkably selective in producing 141b in nearly quantitative yield. The reactor was divided into two stages such that a high degree of conversion of VCl_2 was obtained in the first step. The second stage conditions are adjusted to provide a liquid film on the catalyst. As a result, the yield of 141b is 99.5%, with <500 ppm of VCl_2 remaining in the product. As a result, it is possible to go directly from the reactor into a product tank with little refining. The organic product from the reactor meets the purity specifications defined by the Program for Alternatives Fluorocarbon Testing' (PAFT), an international toxicity panel. PAFT is an industrial consortium of many CFC producers who have shared costs to define toxicity of CFC-alternatives.

Waste Minimization and Resourcing

The ultimate objective of industrial scientists is to develop chemical processes that have 100% yield and operate under conditions that require minimal energy. Unfortunately, modern technology does not allow us to achieve these goals economically. Our challenge is to manage effectively process byproducts and waste through minimization or conversion to salable products (resourcing). Unused products have traditionally been disposed of by incineration, atmospheric venting or underground injection. Historically there has been little incentive to recover and utilize these byproducts. Increasing environmental awareness, disposal costs and concern are now providing the initiative to reconsider these disposal options. A few examples will be given from from DuPont's nylon monomer processes. The main chemistry for the two monomers is shown in Figure 1.

Two moles of hydrogen cyanide are reacted with butadiene, in a two-step process, using a zero valent nickel phosphite catalyst to produce the linear adiponitrile (ADN) and branched methylglutaronitrile (MGN) in very high yield. A small amount of the conjugated, undesirable, 2-pentenenitrile (2PN) has historically been separated and incinerated. The ADN is catalytically hydrogenated to produce hexamethylenediamine (HMD) in very high yield, although small amounts of other cyclic amines such as hexamethyleneimine, (HMI), and 1,2-diamino-cyclohexane are produced. Several years ago we began searching for opportunities to convert these minor byproducts into usable products instead of incinerating them.

The linear, conjugated, 2-pentenenitrile, 2PN, was found to react readily with amines, (HX) to give 2-cyanobutylated amines as shown in equation 2.

$$\underset{2PN}{\diagup\!\!\diagdown\!\!\diagup_{CN}} + HX \longrightarrow \underset{X}{\diagup\!\!\diagdown\!\!\diagup_{CN}} \longrightarrow \underset{X}{\diagup\!\!\diagdown\!\!\diagup\!\!\diagdown^{NH_2}} \qquad (2)$$

Hydrogenation of the aminonitrile with a Raney catalyst leads to a family of branched diamines. Because of the branching, most of the aminonitriles and diamines are liquids at low temperature and have low freezing points. They have found markets as comonomers or curatives, since they lower polymer viscosity, crystallinity, and glass transition temperature.
Catalytic hydrogenation of MGN with a Raney catalyst gives the branched-amine methylpentamethylenediamine, MPMD, and 3-methylpiperidine (3MP) shown in equation 3. The product is dependent on conditions and choice of catalyst. The MPMD was initially isolated from plant streams to develop the market. Many applications were found as a polymer additive in

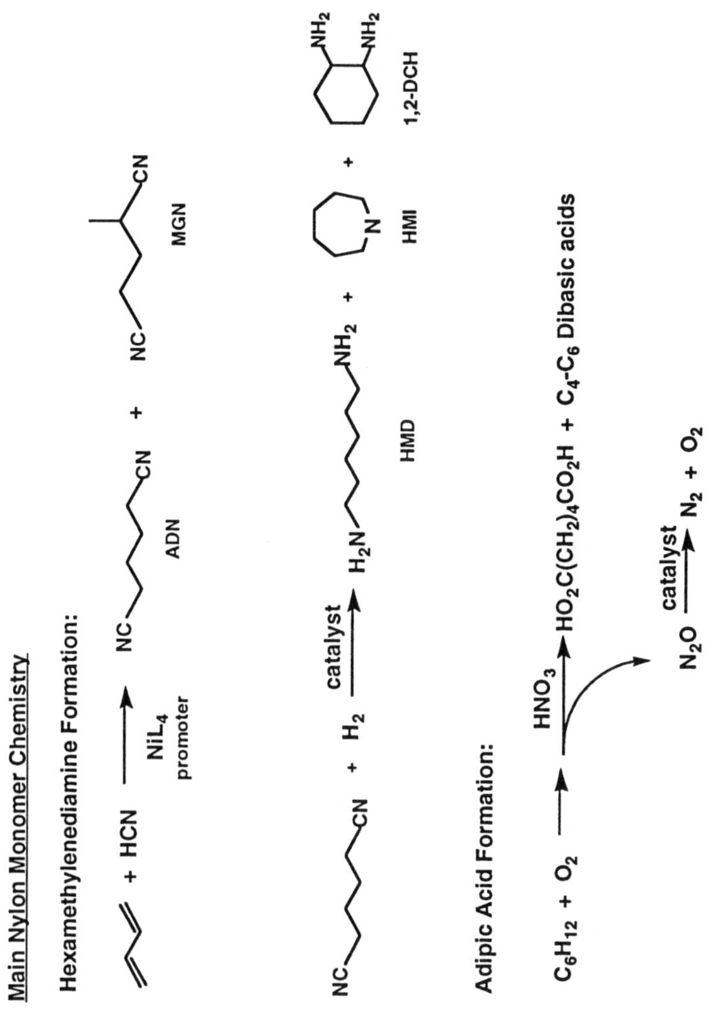

Main Nylon Monomer Chemistry

Hexamethylenediamine Formation:

CH₂=CHCH=CH₂ + HCN →[NiL₄ promoter] NC−(CH₂)₃−CN (ADN) + NC−CH₂CH₂CH(CH₃)−CN (MGN)

NC−(CH₂)₄−CN + H₂ →[catalyst] H₂N−(CH₂)₆−NH₂ (HMD) + hexamethyleneimine (HMI) + 1,2-diaminocyclohexane (1,2-DCH)

Adipic Acid Formation:

$C_6H_{12} + O_2 \xrightarrow{HNO_3} HO_2C(CH_2)_4CO_2H + C_4\text{-}C_6$ Dibasic acids

$N_2O \xrightarrow{catalyst} N_2 + O_2$

Figure 1. DuPont Nylon Monomer Chemistry

urethanes and epoxides to reduce crystallinity and viscosity, as a water treatment chemical and as a monomer in

$$NC\text{-}\!\!\!\diagup\!\!\!\diagdown\!\!\!\diagup\!\!\!\text{-}CN + H_2 \longrightarrow H_2N\text{-}\!\!\!\diagup\!\!\!\diagdown\!\!\!\diagup\!\!\!\text{-}NH_2 + \text{3MP} \quad (3)$$

MGN   MPMD   3MP

polyamides. This material was once incinerated for its fuel value but after a multi-year development effort its use has grown so that a dedicated commercial facility is now used to produce this valuable coproduct intentionally.

Two byproducts, DCH and HMI are produced during ADN hydrogenation (2), the latter resulting from a cyclic deamination reaction. These compounds were initially isolated from plant streams and slowly introduced into the marketplace. Through a long development program, new applications as epoxy-curing agents and agricultural intermediates have been found and developed.

Another example (Figure 1) of waste utilization comes from the other nylon 6.6 monomer, adipic acid. Cyclohexane is oxidized with air to give cyclohexanone and cyclohexanol, which are then oxidized to adipic acid with nitric acid. Although the yields are good, significant quantities of a variety of C_4 to C_6 linear and branched dibasic acids (DBA) are formed. These DBAs were previously burned or disposed of in underground wells. Purification and isolation of the individual DBAs is best done by esterification to dibasic esters (DBEs) and distillation. Many years of market research and development have resulted in a new family of products that are finding wide applications. For example, DBEs are being used as organic solvents for cleaning applications that once used chlorocarbons, thereby addressing two environmental issues with one product. If purified dibasic acids are required, they are easily obtained by hydrolysis of the appropriate ester. A byproduct from the nitric acid catalyzed oxidation of cyclohexanol and cyclohexanone is N_2O which is believed to be an ozone depleting compound. Catalysts are under development for the decomposition of nitrous oxide to N_2 and O_2.

Many other examples from DuPont's polyester and Kevlar processes could be presented to show the clear value of catalytic technology to convert waste byproducts to valuable coproducts. A process that operates at 99% yield with a capacity of one billion lb/year generates 10 million lbs/yr of material that can be burned as a byproduct or converted to a higher value-in-use coproduct. Waste resourcing makes sense if there is a very strong, long term corporate commitment to do so. The examples shown here from DuPont's nylon business, for example, have taken over 10 years. These efforts across the entire coproduct business, currently generate several hundred million dollars in sales. Further efforts are expected to expand this new business considerably by the year 2000, an effort that relies very heavily on catalysis and process R&D.

Asymmetric Catalysis

Everyone is aware of the thalidamide tragedy in which one optically active isomer had good therapeutic effects and the other resulted in birth defects. One study (3) predicts that the trend in manufacture of optically active drugs over the next decade will increase, as shown below:

Figure 2. Trend in Optically Active Drugs

One example of the potential use of asymmetric catalysis comes from our work on asymmetric hydrocyanation (4). The world's fifth largest prescription drug is Naproxen, which is a popular anti-inflammatory compound. Asymmetric hydrocyanation technology offers the potential to produce a precursor to Naproxen, equation 4:

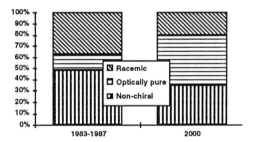

(4)

After considerable research, our scientists decided that chiral carbohydrate ligands (Figure 3) were attractive precursors since their chemistry is well defined. They have several asymmetric centers that can be modified to improve enantiomeric selectivity. Various substituents can be added to change catalytic selectivity and by converting them into phosphinite ligands, steric and electronic effects can be easily studied. A variety of substituents were studied and the effect was quite remarkable. Steric effects were minimal while electron withdrawing groups produced high ee's for the

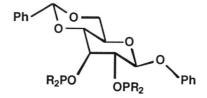

Figure 3. Chiral Phosphinite Ligands for Hydrocyanation

Naproxen precursor nitrile. The best reported results are >90% ee at 25°C. Over 4000 turnovers of catalyst have been shown, without deactivation at 100% conversion. After recrystallization, over 99% optical purity of the nitrile was observed. Although this is only one example, there are many other fine examples of asymmetric catalysis being developed that illustrate the power of catalysis in this area, longer term, particularly for chiral drugs and agrichemicals.

Hazardous and Toxic Materials Management

Hazardous and toxic materials such as HCN, HF, HCl, Cl_2, acrylonitrile, formaldehyde, ethylene oxide, sulfuric acid and phosgene, for example, are essential building reagents in the chemical industry since they often contain functionality or reactivity required for further chemical reactions. Future business practices must avoid or minimize the inventory and transportation of these materials.

Methylisocyanate (MIC) is familiar to us as a result of the tragic incident at Bhopal. It was produced by the phosgenation of methylamine:

$$CH_3NH_2 + COCl_2 \longrightarrow CH_3NCO + HCl \qquad (5)$$

As a consumer of MIC, DuPont was concerned over the use and storage of this toxic material. Prior to Bhopal, we began research on a new process that would produce MIC from less hazardous materials and minimize its handling and storage. The proprietary (5) catalytic oxidative-dehydrogenation process, shown in equations 6-7 was discovered:

$$CH_3NH_2 + CO \longrightarrow CH_3NHCHO \qquad (6)$$

$$CH_3NHCHO + O_2 \longrightarrow CH_3NCO \qquad (7)$$

This innovative research has resulted in a commercial DuPont process that makes MIC and converts it *in-situ* to an agrichemical product. Consequently, the potential for exposure is greatly reduced. This trend in *in-situ* manufacture and derivatization is clearly the way of the future for hazardous chemicals. Non-phosgene routes to isocyanates, and use of solid acids to avoid HF and H_2SO_4 as alkylation catalysts are other examples of research in progress to minimize further the use of hazardous materials.

Environmentally-Safer Products

Although products are introduced into the market to serve societal needs, their impact on the environment is not always predictable. The use of tetraethyl lead in gasoline provided a high octane gasoline for many years. However, lead has now been phased out in certain parts of the world, in favor of environmentally-safer oxygenated organics such as methyl-t-butyl ether (MTBE). New catalytic technology is providing more isobutylene for this large volume chemical. Another recent example (6-7) involves the recognition that ozone is being depleted by man-made chlorocarbons such as methylchloroform, carbon tetrachloride and chlorofluorocarbons (CFCs). These chemicals have served society very well. Atmospheric science, developed during the late 1980's, provided scientific evidence that they were causing significant ozone depletion (nearly 50 years after they were first introduced). As a result, industry has responded rapidly and is currently developing and commercializing safer products. Some of the products are shown in Table I.

Table I. Potential CFC Substitutes

Market	Current CFC	CFC-Alternative
Refrigerants	CFC-12 (CF_2Cl_2)	HFC-134a (CF_3CFH_2)
		HCFC-22 (CHF_2Cl)
		HFC-32 (CH_2F_2)
		HFC-125 (CF_3CF_2H)
		HCFC-124 (CF_3CHFCl)
		HFC-152a (CH_3CHF_2)
Cleaning Agents	CFC-113 ($CF_2ClCFCl_2$)	Blends
Blowing Agents	CFC-11 ($CFCl_3$)	HCFC-141b (CH_3CFCl_2)
		HCFC-123 (CF_3CHCl_2)
		HCFC-22 (CHF_2Cl)

These new products are much more complex than the CFCs they are replacing and require complicated catalytic technology. Although there are many products listed in the table, the main alternative will be HFC-134a. Most CFCs are produced in a single catalytic step from the direct reaction of the chlorocarbon precursor with HF. Temperature and pressure are adjusted to control the degree of fluorination. The alternatives such as HFC-134a can require 2-5 complex catalytic steps as shown in Figure 4 which all add to manufacturing investment, commercialization time and higher cost.

The alternatives are currently acceptable because of their reduced ozone and global warming potentials. The presence of hydrogen results in a lower atmospheric lifetime. The short atmospheric lifetime also results in problems in their synthesis resulting in rapid catalyst deactivation. Longer lived catalysts have subsequently been developed. A detailed summary (8) of the catalytic chemistry reported for the preparation of the most significant CFC alternatives has been reported. Without rapid development of these new catalytic processes, CFCs would continue to be produced, resulting in further ozone depletion.

12. MANZER *Chemistry and Catalysis* 153

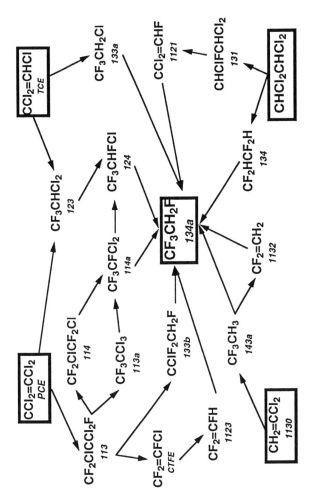

Figure 4. Potential Routes to HFC-134a

As a result of DuPont's efforts to date, over $500 million have been spent.

Conclusions

Catalysis and process research are key technologies for developing environmentally safer processes and products for the future. This will become a condition for staying in business and those companies which commit resources will succeed. This paper has presented a few examples of ways to prevent pollution through waste utilization, new processes with very high yield and new products that are safer to the environment. Further developments can provide an excellent opportunity for collaborations between government, academic and industrial laboratories. The long range research necessary to discover new chemistry and develop mechanistic understandings of current industrial processes can be exciting opportunities for joint collaboration. Much of this work is best done in a university environment where there may be a longer range commitment to research. A major obstacle is often ownership of intellectual property. Universities and industry must find ways to work together to make this process easier.

Literature Cited

1. U.S. Patent (DuPont): 5,105,033 (1992)
2. U.S. Patent (DuPont): 4,937,336 (1990)
3. Chemistry and Industry, 366 (1989)
4. JACS **1992**, *114*, 6265-6266; U.S. Patent (DuPont): 5,175,335(1992)
5. U.S. Patent (DuPont): 4,537,726 (1985)
6. L. E. Manzer, Science, 31-35 (1990)
7. L. E. Manzer, Catalysis Today, *13*, 13-22 (**1992**)
8. L. E. Manzer and V. N. M. Rao, Advances in Catalysis, *39*, 329-350 (**1993**)

RECEIVED August 4, 1994

Tools for Assessment of Benign Chemistry

Chapter 13

Alternative Syntheses and Other Source Reduction Opportunities for Premanufacture Notification Substances at the U.S. Environmental Protection Agency

Carol A. Farris, Harold E. Podall[1], and Paul T. Anastas

U.S. Environmental Protection Agency, Office of Pollution Prevention and Toxics, Mail Code 7406, 401 M Street, Southwest, Washington, DC 20460

> Under section 5 of the Toxic Substances Control Act (TSCA), the U.S. Environmental Protection Agency reviews Premanufacture Notifications (PMNs) for new chemical substances and has authority to limit or ban those substances that may cause an unreasonable risk to human health or the environment. Recently, EPA has begun assessing the pollution potential of the chemical manufacturing processes for certain high volume PMN substances. In parallel with routine PMN review, EPA evaluates feedstocks, solvents, byproducts and impurities to characterize the waste produced and determine whether the generation of toxic substances can be reduced or eliminated. EPA then investigates possible approaches to pollution prevention, including improved and alternative syntheses. If EPA identifies potential pollution prevention opportunities, it corresponds with the PMN submitter to suggest voluntary consideration of EPA's pollution prevention approach. This paper explains EPA's assessment methodology and results to date.

Each year over a thousand new chemical substances are manufactured or imported in the United States. While some of these substances will replace other more toxic substances, each new substance has the potential to increase the levels of pollution in our air, water and land. Even substances that are themselves innocuous may be manufactured using highly toxic feedstocks or may result in significant generation of hazardous waste. The U.S. Environmental Protection Agency has regulatory authority over new chemical substances as well as the mandate to champion pollution prevention in all aspects of our environment.

---

[1]Current address: U.S. Environmental Protection Agency, Office of Pesticide Programs, Crystal Station, Arlington, VA 22202

A major responsibility of the Office of Pollution Prevention and Toxics is the implementation of the Toxic Substances Control Act (TSCA) (*1*). Central to TSCA is section 5, which requires manufacturers to submit premanufacture notifications (PMNs) to EPA before they begin manufacture of new chemical substances. Under section 5, EPA may limit or ban the production of a new chemical substance that may present an unreasonable risk to human health or the environment.

The PMN review program has been in place at EPA since 1979 and each year the Agency staff review approximately 2,000 new PMN cases. In the past, the Agency's review has focused almost exclusively on the PMN substance itself and not on the feedstocks, process streams or any wastes that might be generated. Now, however, EPA chemists are also reviewing these related substances in a new process that runs parallel to routine PMN review. This paper discusses that process.

EPA's Office of Pollution Prevention and Toxics (OPPT) is charged with implementing both the Pollution Prevention Act of 1990 (PPA) (*2*) and TSCA. The PPA states that "The Congress hereby declares it to be a national policy that pollution should be prevented or reduced at the source whenever feasible... ." Source reduction is defined, in part, as any practice that "reduces the amount of any hazardous substance, pollutant or contaminant entering any waste stream prior to recycling, treatment, or disposal; ... ." Although the PPA does not specifically mandate that EPA consider pollution prevention for PMN cases submitted to it under TSCA, EPA chemists thought it possible to identify significant opportunities for source reduction for those toxic substances associated with the manufacture of PMN substances.

During a source reduction (or pollution prevention) review, the Agency would identify potential waste in the synthetic process or use of raw materials associated with PMN cases. Among the wastes of interest are excess or unconverted reactants that are not recovered or recycled, side reactions, excess reagents used for separation or recovery of the product and solvents or catalysts that are not recovered or recycled. Having identified these wastes, EPA chemists would attempt to develop suggestions to reduce or eliminate them. The aim was a program to notify PMN submitters of pollution prevention opportunities for their voluntary consideration. With the increasing costs of waste treatment, waste disposal, regulatory compliance and liability, many industries are finding that pollution prevention can also lower their operating costs. Ultimately, the Agency hopes that such considerations by manufacturers will reduce pollution, increase manufacturing efficiency and make good economic sense.

Requirements for Source Reduction Reviews

The Agency had three requirements for setting up a source reduction review: sufficient information for the review, a workable assessment methodology and the ability to propose practical solutions to waste problems.

PMN Information. PMN submissions must be made on a standardized Agency form (*3*); the information must be supplied by PMN submitters to the extent that it is known or reasonably ascertainable. The information requested by the form is mandatory with the important exception that manufacturing information, including the process flow diagram, worker exposures and batch size and number, is required for domestic manufacture but not for substances that are imported to the U.S. The form also contains a page for optional pollution prevention information.

The types of information contained in PMNs that are useful for source reduction assessments include:

PMN substance identity;

a process flow diagram including the identities and amounts of feedstocks used, and the major unit operations and chemical conversion steps;

the amount of product produced per batch or per unit time (if the process is continuous);

the number of batches per year (or the operating time per year);

the expected annual production volume;

the identity of impurities and byproducts (a byproduct is defined for our purposes as "a chemical substance produced without separate commercial intent during the manufacture or processing of another chemical substance or mixture"); and

any optional information submitted on the form's pollution prevention page.

The optional pollution prevention page is extremely useful to EPA. Pollution prevention technologies planned by manufacturers may be included in the PMN submission so that the EPA reviewers can take them into account when assessing the risk of the substance relative to others already in commerce. These pollution prevention considerations also affect EPA's regulatory determinations. If PMN submitters have thoroughly considered the pollution prevention possibilities for their synthesis, it may be unnecessary for EPA to do any further pollution prevention review.

The amount of information required with PMNs appeared to be sufficient for EPA chemists to derive a rough material balance and obtain identities and levels of wastes involved in the process. A brief review of 50 PMN submissions indicated that at least 90% appeared to have sufficient information for this analysis.

Other Requirements. Success of source reduction reviews depended upon a reasonable assessment methodology that would identify the most wasteful processes and develop suggestions to improve them without requiring large resource expenditures on the Agency's part. A basic premise was that chemists could identify quickly those PMN cases with the greatest waste so that intensive review and development of options for reducing the wastes would only be done on the most appropriate cases.

The last requirement was for EPA chemists to develop solutions to waste problems that would be reasonable and technologically feasible; the analysis of economic feasibility and other suitability would, of course, be left to the submitter. EPA chemists expected to draw upon their years of experience assessing PMN syntheses including, more recently, pollution prevention measures submitted with PMNs. An assumption here was that PMN submitters may not necessarily have done such source reduction analyses themselves and also may not be aware of non-proprietary technologies from other industry sectors that EPA also regulates.

Pilot Study for Source Reduction Reviews

In February 1991, EPA chemists initiated a pilot study to test the assessment process and to determine whether source reduction reviews were feasible. EPA chemists designed a process to concentrate resources on those cases most likely to yield significant source reduction opportunities. The process began with a quick sort to determine which PMN cases had sufficient potential for further study. This was followed by a preliminary assessment to identify those cases with the highest pollution prevention potential and then a detailed assessment during which pollution prevention options were developed and analyzed. Finally, the Agency determined whether there were source reduction suggestions of sufficient merit to justify a letter to the PMN submitter.

The pilot study included 418 consecutive PMN submissions. Of those cases, 11% received preliminary assessments. Of these cases, Agency chemists were able to identify source reduction opportunities for 13% of the cases that went into preliminary review. It is likely that this percentage will increase in the future as a greater number of new pollution technologies become available. In addition, the Agency determined that the resources to perform these assessments were not particularly great, so the program appeared to be feasible from an economic standpoint.

Ongoing Source Reduction Assessments

EPA staff currently perform source reduction assessments on all new PMN cases in a process called $\underline{S}$ynthetic $\underline{M}$ethodology $\underline{A}$ssessment for $\underline{R}$eduction $\underline{T}$echnologies (SMART). Although minor changes in the review process and criteria have been instituted since the pilot study, the following process description reflects the Agency's current process and criteria.

Screen for Preliminary Assessment. The Agency uses four criteria for its first screen. If a PMN case meets all four criteria, a preliminary source reduction assessment is done.

First, the PMN must have a projected production volume equal to or greater than 25,000 kg/yr. This cut-off point focuses our attention on those cases that may have the greatest potential for pollution due to their volume and also the greatest potential for significant pollution prevention. In the future, the Agency may lower the cut-off volume to allow for review of more cases.

Second, the PMN must include a process description or process flow diagram. PMNs meeting this criterion are for substances manufactured in the U.S. rather than imported, since only PMNs for domestic manufacture are required to have process information.

Third, the submission must be a full PMN, as opposed to an abbreviated exemption notice. Notices for test market exemptions, low volume exemptions or polymer exemptions are dropped from the review at this point. The three types of exemption notices require quick-turnaround reviews based on a presumption of low exposure or low risk. Test market exemption substances are produced in limited quantities for a limited time to determine their marketability prior to PMN submission. Low volume exemption substances fall under our 25,000 kg/yr cut-off. Substances eligible for the Agency's polymer exemption are a group of especially low-risk polymers and are presumed to be poor candidates for pollution prevention. Because exemptions are only for categories of new chemical substances that "will not present an unreasonable risk to human health or the environment," (1) exemption substances are unlikely to yield significant source reduction opportunities.

Fourth and last, the PMN substance must not be a polymer or complex reaction mixture. Polymers were excluded because their synthetic reactions tend to incorporate virtually all of their feedstocks into product. Complex reaction mixtures were excluded because many do not lend themselves to rapid assessment; in the future, the Agency may decide to include them if the pollution prevention benefits are likely to warrant the greater expenditure of resources.

Preliminary Assessment. The first step in preliminary assessment is to determine the sources, identities and amounts of the waste products generated in the manufacturing process, that is to perform a material balance. This is done without consideration of any waste treatment or control technology. Generally, the amounts of waste are calculated based on the production volume anticipated by the PMN submitter. In most cases, the information in the PMN will be sufficient for these calculations; if not, the EPA chemist may contact the submitter for clarification or additional information.

Next, the toxicity of each individual waste is estimated and the level of EPA's concern for that waste is determined. The toxicity estimation is done by comparing the wastes with previously-published lists of substances.

For the exclusive purposes of this review process, the Industrial Chemistry Branch (ICB) of OPPT has developed a short list of *extremely toxic substances*, for which there is concern at very low levels. This list is found in Table I. This list is subject to change as additional data become available.

Table I. Extremely Toxic Substances

Hydrogen fluoride	Arsenic compounds
Hydrogen sulfide	Beryllium compounds
Organomercurials	Cadmium compounds
Phosgene	
Dioxins/Benzofurans	
-- all Cl and Br isomers with 4-8 halogens	
Certain highly neurotoxic organophosphorus compounds	
e.g., Sarin	

At the other end of the spectrum, ICB also has a short list of *relatively innocuous substances*. Agency staff were conservative in placing substances on this list. Substances for which there is some uncertainty as to whether they should be in this category are placed in the third category, potentially hazardous substances. The list of relatively innocuous substances is found in Table II; this list is also subject to change.

Table II. Relatively Innocuous Substances

Water	Na, K, Mg or Ca chloride
Nitrogen	Na, K, Mg or Ca carbonate
Carbon dioxide	Na, K, Mg or Ca sulfate
Certain carbohydrates in solution	
e.g., Starch	
Sucrose	
Certain high molecular weight condensed-phase polymers	
e.g., Polyethylene	
Polyethylene terephthalate	

All other substances are classified as *potentially hazardous*; this broad middle category includes substances with a wide spectrum of toxicity, but it serves the Agency's screening needs.

During the next step, EPA chemists use two lists of substances to classify the wastes as being *wastes of concern* or not. These lists are: the OPPT New Chemicals Program Categories of Concern for PMN substances (*4*) and the Superfund Amendments and Reauthorization Act (SARA) (*5*) section 313 Toxic Chemicals (the Toxic Release Inventory or TRI) (*6*). Any PMN waste found on either of these lists is considered to be a waste of concern for the next step in the process. The Categories of Concern is an evolving list developed to facilitate PMN review; PMN substances that fit within the categories of concern may be subject to Agency regulation unless there is mitigating information or test data provided with the PMN. Substances on the TRI list are subject to reporting under Title III of SARA, the Emergency Planning and Community Right-to-Know Act. This list is also expected to evolve over time.

Selection for Detailed Assessment. A PMN case meeting one or more of the five criteria below is given a detailed Agency assessment for source reduction. PMN cases that do not meet any of the criteria are dropped from further pollution prevention review. A detailed assessment is performed if:

(1) an *extremely toxic* substance is present anywhere in the process stream, in any amount. The Agency would like to find ways to avoid the use of any of these

substances; PMN cases associated with any of the extremely toxic substances therefore are given detailed assessments.

(2) any solvent that is *potentially hazardous* is not recycled or if a coproduct is generated that is *potentially hazardous* and is not commercialized or used captively. (Recall that potentially hazardous substances are those in our broad middle category of toxicity.) Coproducts are defined as substances necessarily produced in the same reaction in which the PMN substance is produced; the term includes byproducts. This criterion identifies wasteful processes that might be amenable to source reduction.

(3) the *waste of concern* includes a substance on the TRI list that is greater than or equal to the threshold volume for TRI reporting. That threshold is 11,340 kg/yr (i.e., 25,000 lb/yr). Any substance generated during the production of a PMN substance that would trigger reporting under TRI is of sufficient concern to merit a detailed assessment.

(4) the total amount of *waste of concern* exceeds the trigger level set for that production volume. The use of trigger levels is simply a screening guide to identify the most wasteful processes and not an Agency position that any particular level of pollutant is acceptable. The *waste of concern* is expressed as a weight percentage of the PMN chemical production and then is compared with the trigger level set for that production volume, expressed as percent waste. Waste in excess of this maximum level triggers a detailed assessment. These trigger levels are given in Table III.

Table III. Trigger Levels for Wastes of Concern

Production Volume (PV) of PMN Substance (kg/yr)	Trigger Level (% Waste)
25,000 ≤ PV < 250,000	10
250,000 ≤ PV < 500,000	8
500,000 ≤ PV < 750,000	6
750,000 ≤ PV < 1,000,000	4

At production volumes at or in excess of 1,000,000 kg/yr, the trigger level is set at 20,000 kg/yr of waste of concern.

(5) the total amount of *potentially hazardous waste* exceeds the trigger level set for that type of waste. The trigger level is calculated from the production volume of the PMN substance using equation 1.

$$y = 77 - 12\log(x) \tag{1}$$

where x = production volume of the PMN substance in kg/yr (for volumes below 1,000,000 kg/yr) and y = the trigger level expressed as percent waste. For simplicity, calculated trigger levels are rounded to the nearest whole number. This equation generates a curve with approximately 25% as the trigger level for 25,000 kg/yr production volume to 5% as the trigger level for 1,000,000 kg/yr. For all production volumes at or above 1,000,000 kg/yr, the trigger level is 50,000 kg of waste.

For both *wastes of concern* and *potentially hazardous wastes*, the decrease in the percentage of waste with increasing production volume reflects expected increases in process efficiencies at higher production volumes. Of course, at high production volumes, even low percentages of waste account for large amounts in absolute terms. For this reason, the trigger levels were chosen to be absolute numbers above 1,000,000 kg/yr. This is only a simple screening tool to identify the most wasteful processes and not an Agency position that any particular level of pollutant is acceptable.

Detailed Assessment. During detailed assessment of the PMN cases that meet the criteria to continue in the process, more careful calculations of the wastes may be made, if necessary. Most of the Agency's effort, however, is concentrated on developing potential solutions to the problems identified during the previous review. EPA chemists look for creative solutions based on their general knowledge of synthetic organic chemistry and their experience reviewing syntheses for thousands of new chemical substances across many industrial sectors. The Agency may also work with the PMN submitter's chemists to look for potential improvements. Agency chemists are looking forward to using a variety of tools, including computer synthesis programs to help find better syntheses for PMN substances during detailed assessment.

Where reasonable pollution-preventing suggestions are identified that are likely to make a difference in the level and type of waste, the Agency summarizes its findings and suggestions in letters for the voluntary consideration of submitters. In sending these letters, the Agency is asking the PMN submitter to consider a change in its synthesis, if such proves feasible technically and economically. Because PMNs are submitted prior to the first manufacture of the new substances for commercial purposes, it may be easier of a manufacturer to make changes to its synthetic process than it would be for substances that are already established in the marketplace.

Examples of Results to Date

The Agency's requirements to protect all claims of confidential business information in PMNs precludes a detailed description of the source reduction suggestions that have been forwarded to submitters to date. Some general examples may, however, be made public.

In one recent case for an aliphatic amine, three concerns were identified by EPA: the use of a feedstock that was a *waste of concern*, a *potentially hazardous* byproduct and the loss of nearly 6% of the final PMN product during the course of the manufacturing process. Using the Agency's suggestions, the PMN submitter was able to reduce the predicted amount of PMN substance loss in water waste by 98.5%. This resulted in thousands of kilograms of the PMN substance being available for its intended use rather than being released to the environment.

In other cases, the Agency has not received word from the submitter as to whether the pollution prevention suggestions in our letters have had an effect on the manufacturing process. In some of these cases, commercial production may not have begun; in others, the submitter may not have communicated with us after evaluating our suggestions or the Agency's suggestions may not have been suitable for a variety of unforeseen economic or technical reasons. Agency suggestions illustrating the letters sent to date include the following:
(1) A new synthesis as an alternative to a chlorination reaction with excess chlorine. Our suggestion involved a substitute feedstock that is readily available. Further, the Agency encouraged the submitter to investigate this and other substitute feedstocks.
(2) A modification of reaction conditions that might decrease the formation of a toxic byproduct; further, more of the PMN substance could possibly be recovered from waste water.
(3) An amine feedstock might be reclaimed and re-used rather than incinerated, as was planned in the PMN submission.
(4) Changing the stoichiometry of a chemical reaction to reduce the amount of excess feedstock required to drive the reaction.
(5) More efficient uses of solvents (several different letters).
In all cases, the Agency encourages PMN submitters to discuss the feasibility of our suggestions with us following their laboratory evaluation; this feedback will enable the Agency to strengthen its pollution prevention analyses.

Discussion and Conclusions

We have shown that source reduction assessments are feasible as a part of routine PMN review at EPA. Submitters may not have thought about the potential for pollution prevention that their new manufacturing operation affords; yet, with costs to dispose of hazardous materials rising and end-of-pipe restrictions increasing, source reduction becomes more cost-effective. Eventually, the Agency would hope that PMN submitters would have performed an independent analysis of the pollution prevention options that might be open to them such that the process described in this paper would not be necessary. Submitters would include the results of their analysis on the PMN pollution prevention page for Agency consideration as part of their submissions. However, with the unique perspective that Agency chemists have from their oversight of all new substances and their focus on pollution prevention, it appears that the Agency's review will continue to be useful for some time.

EPA presents suggestions to PMN submitters, who ultimately will determine whether changes are feasible. If, however, as a result of these reviews, the Agency encourages submitters in general to think more about source reduction (including alternative syntheses), and, if some of the suggestions are in fact adopted, the Agency will have been successful in promoting one important aspect of pollution prevention.

Acknowledgments

The authors wish to acknowledge the suggestions offered by Drs. Roger Garrett, Steven Hassur and Tracy Williamson of the Industrial Chemistry Branch, Office of Pollution Prevention and Toxics, EPA in the development of the assessment methodology and the assistance of Drs. Greg Fritz and Daniel Lin of Technical Resources, Incorporated, Washington, D.C. in performing the technical review for the Pilot Study.

Disclaimer

This chapter was prepared by Harold Podall, Paul Anastas and Carol Farris in their private capacity. No official support or endorsement of the U.S. Environmental Protection Agency is intended or should be inferred.

Literature Cited

1. Toxic Substances Control Act. 15 U.S.C. §§2601-2629, **1982** and Supp. III **1985**.
2. Pollution Prevention Act of 1990. 42 U.S.C. §§13101-13109, **1990**.
3. U.S. Environmental Protection Agency. 40 CFR 720 and EPA Form 7710-25. Premanufacture notification.
4. OPPT New Chemicals Program Categories of Concern (under TSCA). A publication of the U.S. EPA Office of Pollution Prevention and Toxics, updated periodically and available from the TSCA Assistance Information Service, (202) 554-1404; (202) 554-0551 (TTD); (202) 554-5603 (on-line service modem).
5. Emergency Planning and Community Right-to-Know Act (also known as Title III of the Superfund Amendments and Reauthorization Act). 42 U.S.C.A. §§11001-11050. **1986**. Section 313. 42 U.S.C.A. §11023.
6. Superfund Amendments and Reauthorization Act section 313 Toxic Chemicals (Toxic Release Inventory list). Available from the Emergency Planning and Community Right-to-Know (Title HI of SARA) Hotline, (800) 535-0202 (national) or (703) 412-9877 (Virginia).

RECEIVED July 26, 1994

Chapter 14

Computer-Assisted Alternative Synthetic Design for Pollution Prevention at the U.S. Environmental Protection Agency

Paul T. Anastas[1], J. Dirk Nies[2,3], and Stephen C. DeVito[1]

[1]U.S. Environmental Protection Agency, Office of Pollution Prevention and Toxics, Mail Code 7406, 401 M Street, Southwest, Washington, DC 20460
[2]Dynamac Corporation, Rockville, MD 20850–3268

> The usefulness of computer assisted organic synthesis (CAOS) in identifying alternative, potentially more environmentally benign reaction pathways for the synthesis of commercial chemicals was explored. Software programs for synthetic and retrosynthetic design were identified from the scientific literature and through contact with experts in CAOS. Three programs (CAMEO, LHASA, and SYNGEN) were obtained and evaluated on their (1) ability to generate synthetic pathways that are consistent with methods reported in the literature, (2) ability to propose chemically useful alternatives to established synthetic pathways, and (3) ease of use. Our preliminary evaluation indicated that these computer programs possess intriguing potential for proposing alternative reaction pathways that may subsequently be evaluated for their relative risk and economical viability.

People have long recognized the many benefits that have been achieved through chemistry and its application to fields such as agriculture, medicine, electronics, textiles, and transportation, to name a few. However, as a society we have learned from experience that certain chemicals believed to be innocuous may manifest toxic effects only following years of wide-spread exposure to the general population. In fact, deliberate release to the environment of chemicals whose toxicity was well known went unchallenged and unregulated for many years. In an attempt to minimize exposure to hazardous chemicals, Congress and state legislatures passed laws that require hazardous waste treatment or that restrict environmental release. These "end-of-pipe" approaches to controlling pollution generally have been successful in reducing releases of toxic substances into the environment, but less than originally anticipated. Hazardous waste disposal sites, for example, are designed to contain hazardous substances permanently so that they do not enter the environment. Over the years, however, many have begun to leak and have, ironically, become major sources of environmental contamination. In addition, the cost of dealing with pollutants after their generation is becoming prohibitively expensive. Perhaps most importantly, the end-of-

---

[3]Current address: Chemical Information Services, Inc., Rockville, MD 20850–4211

pipe approach to controlling pollution does not prevent the generation of pollution; its focus is to reduce exposure to pollution *after* it has been created.

There has emerged in recent years a new environmental paradigm known as pollution prevention. The underlying philosophy of this paradigm is fundamentally simple: we must avoid creating pollution whenever possible. The pollution prevention paradigm is very much like the preventative medicine paradigm: preventative medicine focuses on preventing illnesses from occurring, rather than finding cures and treating illnesses after they have occurred, whereas pollution prevention focuses on preventing the creation of pollution, so as not to have to contend with the arduous task of dealing with it later. In 1990 the United States Congress passed the Pollution Prevention Act, which requires chemical companies to prevent, whenever possible, the creation of pollution (*1*). Pollution prevention has become the foremost priority of the United States Environmental Protection Agency (EPA), and many of EPA's programs have been restructured so as to incorporate pollution prevention considerations in risk assessment and risk management decisions (*2*).

Alternative Synthetic Design

Organic synthesis is the art of making and isolating a particular organic substance from the reaction(s) of other organic substances, from which a myriad of organic products theoretically may form. In learning the basic skills of synthetic organic chemistry, graduate students primarily have been trained to make and to isolate a desired target substance with little emphasis on the toxic nature of reagents, solvents, or byproducts used or associated with the synthesis of the target substance. Traditionally, in large-scale production of organic substances, the commercial organic chemist not only was challenged with the often difficult task of designing a synthetic route to a target chemical, but the synthesis had to provide the greatest yield of desired product at the lowest direct cost (indirect costs such as those associated with pollution abatement, worker protection, hazardous waste treatment and disposal typically were not considered). Because product yield and direct cost were usually the primary concerns in synthetic design of commercial chemicals, little regard was given to the toxic nature of substances used in (or created from) commercial organic synthesis.

A major source of pollution occurs from chemical syntheses. Many starting reagents, solvents, or products from side reactions are quite toxic, and contribute greatly to overall pollution. Logically, designing alternative syntheses such that they do not require toxic starting reagents (or solvents) or produce toxic byproducts should have a major impact in preventing pollution at the source. Creating pollution in synthetic organic chemistry (whether at a student's lab bench or in commercial manufacture) traditionally has not been a concern in synthetic design. With the recent emphasis on pollution prevention, however, the synthetic organic chemist has a new challenge in planning synthetic strategies: to design organic syntheses oriented toward preventing pollution. The modern day organic chemist must consider the environmental impact of a chemical reaction; product yield and direct cost are no longer the only considerations. Students of organic synthesis similarly must learn to structure their thinking towards the environmental consequences of making chemicals if they are to be prepared for what will be expected of them after they have completed their academic training. In fact, many academic researchers have been remarkably successful in identifying less polluting alternative synthetic reactions for making commercial chemicals. Draths, et al., for example, recently have reported an alternative synthesis for making hydroquinone (a high production volume commercial chemical) that obviates the need for aniline and benzene, which are quite toxic and currently used in the commercial manufacture of hydroquinone (*3*). In the new (alternative) synthesis, hydroquinone is produced enzymatically from D-glucose. Other examples of academic research in identifying environmentally benign alternative syntheses have recently been summarized (4), some of which are described elsewhere in this book.

Computer Assisted Alternative Synthetic Design. Within the past two decades there have been enormous advancements in computer technology. The emergence of software programs that not only retrieve and analyze large quantities of data but also perform tasks that require strategy and tactics ("intelligence") have had significant impacts on many areas of technology. Beginning in 1967, E.J. Corey and his colleagues at Harvard University began to design and build computer programs for planning syntheses of organic chemicals. Since that time, at least 45 significant synthesis programs have been developed, and more are under development. These synthesis programs are intended to assist the organic chemist in designing synthetic pathways to target molecules. Using computer assisted organic synthesis (CAOS), organic chemists are now able to identify reaction pathways that otherwise may not have come to mind.

Most of the synthesis software programs available at present generate syntheses for target molecules by working backwards (retrosynthetically) from the target molecule to candidate starting materials. Other programs work in the foward direction (synthetically) in that they require starting reagents and reaction conditions as input, from which theoretical reaction product(s) are generated. Programs which work retrosynthetically are useful for identifying theoretical synthetic pathways, whereas programs that work in the forward direction (synthetically) can identify products, side reactions, by-products, and the effects of varying reaction conditions. In theory, concomitant application of retrosynthetic and synthetic programs permits optimal routes to be identified, and their associated conditions, by-products, estimated costs and potential hazards to be compared. It is important to stress that presently available CAOS software programs were not written to incorporate pollution prevention strategies in synthesis design. Users of currently available CAOS software programs, however, may be able to identify syntheses that appear less polluting than other syntheses. Our ultimate goal is to use CAOS software programs to help us identify less polluting syntheses that we can suggest to chemical manufacturers as possible alternatives to syntheses known to create significant quantities of pollution. We were interested in exploring the usefulness of CAOS software programs at EPA for identifying alternative reaction pathways that we feel are environmentally advantageous to the reaction pathways encountered, for example, in EPA's Premanufacture Notification (new chemicals) program and existing chemicals program. The purpose of this study was to assess the reliability of some of the currently available CAOS programs in predicting syntheses of known substances, and the products and byproducts of known reactions.

Methods

A search of the scientific literature for articles and reports that describe synthetic design and related software programs and applications was conducted. A list of the CAOS-related articles is provided in the Literature Cited section (5-67). In addition, academic and industry experts in the U.S. were contacted to obtain information regarding their systems and to inquire if they are aware of other potentially useful software packages. More than 45 different synthetic design software systems were identified.

Selection of software programs for evaluation was based on: (1) applicability of software to a broad range of reaction types and organic substances that are representative of those reviewed at EPA; (2) willingness of authors to provide evaluation copies of their software; and (3) compatability with EPA hardware. Using these criteria three synthesis programs were identified and obtained: Computer Assisted Mechanistic Evaluation of Organic reactions (CAMEO, 1992 version); Logic and

Heuristics Applied to Synthetic Analysis (LHASA, version 12.1) and Synthesis Generator (SYNGEN, version 2.3, *15*).

These three programs use different methods to translate chemical knowledge into synthetic proposals. CAMEO works in the foward or "synthetic" direction, requiring starting reagents as input, and computes a product (or products) by applying a series of rules designed to consider structural features to determine reactivity. LHASA and SYNGEN work retrosynthetically, requiring a target molecule as input and computing starting reagents as output. LHASA draws upon a knowledge base of more than 1,000 reaction transforms to suggest precursors that are one step removed from the target. SYNGEN uses mathematical methods to store structures, to describe chemical reactions, and to generate all possible precursors (within the constraints of its library of reactants) in one operation.

Each program was evaluated on the following: (1) ability to generate synthetic pathways of a particular compound that are similar to those reported in the literature; (2) ability to generate rational alternative syntheses to established synthetic pathways; and (3) ease of use and user-friendliness. CAMEO, LHASA, and SYNGEN were loaded on a VAX computer at the EPA's National Computing Center (NCC), Research Triangle Park, North Carolina, and evaluated using a personal computer equipped with a mouse. Access to the VAX computer was achieved using TEEMTALK telecommunications and Tektronix terminal emulation software. Selection of target molecules and reaction pathways were based on both commercial interest and theoretical interest. More than 50 reactions and target molecules were examined; this paper presents a few representative examples.

The synthetic ability of CAMEO was evaluated using a number of known reactions, including the reactions of:

1) methyl isocyanate with 1-naphthol, the commercial synthesis of carbaryl (i.e., 1-naphthyl-N-methylcarbamate) (*68*);

2) ethyl diazoacetate with 4-vinylpyridine (the known reaction for the synthesis of 4-(2-carbethoxycyclopropyl)pyridine) (*69*);

3) 1,3-cyclohexadiene with maleic anhydride (a known synthesis of *endo*-bicyclo[2.2.2]oct-5-ene-2,3-dicarboxylic anhydride) (*70*);

4) thiophene with benzoyl chloride (a known synthesis of 2-benzoylthiophenene) (*71*); and

5) N-(2-phenylethyl)acetamide in the presence of a Lewis acid and heat (the known synthesis of 1-methyl-3,4-dihydroisoquinoline) (*72*).

The retrosynthetic ability of LHASA was evaluated by comparison with known syntheses of a number of compunds including: 1) carbaryl (*68*); 2) 1-methyl-3,4-dihydroisoquinoline (*72*); and 3) quinoline (*73*).

The retrosynthetic ability of SYNGEN was evaluated similarly to that of LHASA including the compounds 1) carbaryl (*68*); 2) 4-(2-carbethoxycyclopropyl)-pyridine (*69*); 3) 5,8,9,10-tetrahydro-1,4-naphthoquinone (*70*); and 4) 4-nitrodiphenylamine (*78*).

Results

CAMEO Assessment. Using starting materials and reaction conditions as input, CAMEO predicts the products of organic reactions using a mechanistic selection of algorithms for a wide variety of reaction modules (i.e., Carbenoid, Basic/Nucleophilic,

Acidic/Electrophilic, Electrophilic Aromatic, Radical, Heterocyclic, Oxidative/Reductive, and Pericyclic) that are representative of fundamental organic reaction types. CAMEO perceives rings, functional groups, potential electrophilic and nucleophilic sites, aromaticity, and stereochemistry to determine reactivity. CAMEO also estimates pKa values for acidic hydrogens and checks for chemical instability and structural strain. Erroneous structures are returned to the user for correction. CAMEO then combines the reactants following the mechanistic rules associated with the reaction module selected. The heat of reaction is estimated and a product ranking scheme designates products as "major", "minor", or "disfavored." A "tree" menu is available to display the structure and relationship between reactants, intermediates, and products. Detailed comments regarding the mechanistic choices made by CAMEO are available to the user. To explore all potential reactions and the products of these reactions fully the user must separately evaluate a given set of starting materials and reaction conditions under several if not all of the available reaction modules.

Ease of Use and User Friendliness. Once a chemical structure has been entered, the user must select a specific mechanistic module (i.e., Carbenoid, Basic/Nucleophilic, Acidic/Electrophilic, Electrophilic Aromatic, Radical, Heterocyclic, Oxidative/Reductive, or Pericyclic) for evaluation of the reactants and define the reaction conditions, reagent, and solvent. Then the user instructs CAMEO to run the reaction. Within approximately 10 seconds, the results are displayed on the screen.

The menu screens of CAMEO are well designed and easy to follow, and enable easy operation of the program.

Synthesis Predictions. The well-known reaction of methyl isocyanate (**1**) with 1-naphthol (**2**) to yield carbaryl (**3**) *(68)* (Scheme 1) was correctly predicted by CAMEO.

Scheme 1. Synthesis of Carbaryl (3); Reaction of Methyl Isocyanate (1) with 1-Naphthol (2)

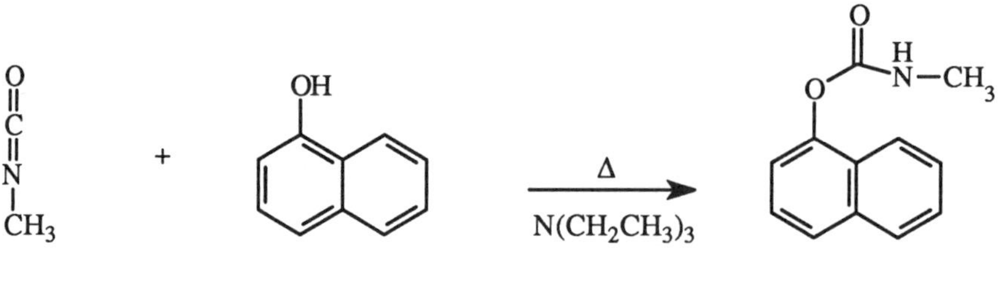

No by-products were identified by CAMEO and in this case none were expected. CAMEO predicted carbaryl as the reaction product, however, only if the Acidic/Electrophilic mechanistic module of the program was selected. No reaction was found if the nucleophilic module was chosen, which would have appeared to be the obvious choice.

The reaction of ethyl diazoacetate (**4**) with 4-vinylpyridine (**5**) to yield ethyl 4-(2-carbethoxycyclopropyl)pyridine (**6**) *(69)* (Scheme 2) was correctly predicted by CAMEO using the Carbenoid module; however, this reaction, which was reported in the literature to proceed when heated, was found to "proceed" in CAMEO only when ultraviolet light was selected to initiate the reaction. In addition, an unusual and seemingly unlikely 4-vinyldihydropyridine product (**7**) was also predicted (Scheme 2).

Scheme 2. Synthesis of 4-(2-Carbethoxycyclopropyl)pyridine (6); Reaction of Ethyl Diazoacetate (4) with 4-Vinylpyridine (5)

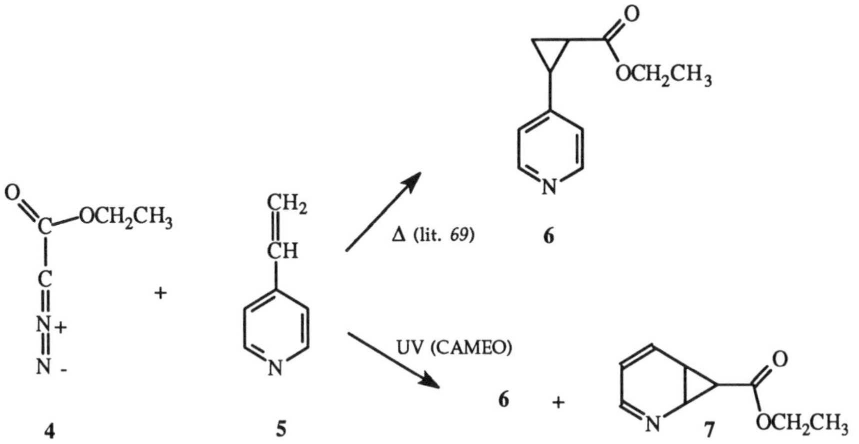

Using the Pericyclic module, the Diels-Alder cycloaddition reaction of 1,3-cyclohexadiene (8) with maleic anhydride (9) (a known synthesis of *endo*-bicyclo[2.2.2]oct-5-ene-2,3-dicarboxylic anhydride, 10 *(70)*), was properly predicted by CAMEO, but only as a minor product (Scheme 3). CAMEO predicted the 3-(1,4-cyclohexadiene) succinic anhydride (11) adduct as the major product via a "forward ene" mechanism.

Scheme 3. Synthesis of *endo*-Bicyclo[2.2.2]oct-5-ene-2,3-dicarboxylic Anhydride (10); Reaction of 1,3 Cyclohexadiene (8) with Maleic Anhydride (9)

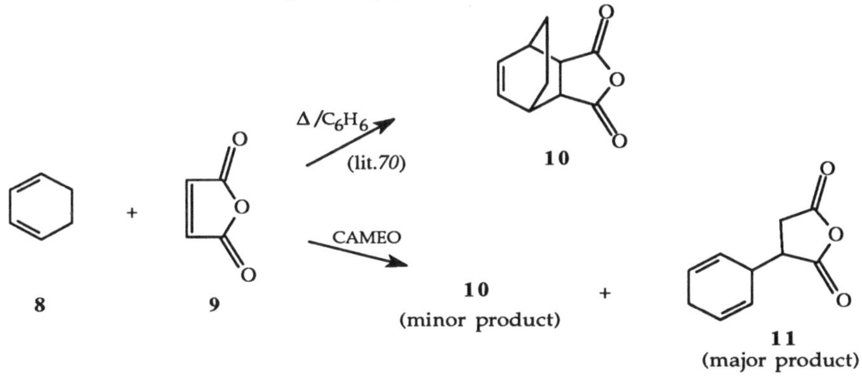

CAMEO failed to predict the formation of 2-benzoylthiophene (12) via Friedel-Crafts acylation of thiophene (13) with benzoyl chloride (14) in the presence of a Lewis acid (e.g., stannic chloride) using the Electrophilic Aromatic module. Thiophene is highly reactive under these conditions and would have been expected to undergo acylation readily, as previously reported *(71)* (Scheme 4). Instead, CAMEO predicted that stannic chloride (the catalyst) would replace a hydrogen in either the 2- or 3-position of

172 BENIGN BY DESIGN

thiophene to form a tin-carbon sigma bond (15) (Scheme 4). The Basic/Nucleophilic module of CAMEO predicted nucleophilic substitution in which sulfur displaced chlorine to yield the phenyl thiophenium ketone (16). The Acidic/Electrophilic module predicted the reaction of the benzoyl carbonyl carbon across the thiophene double bond (17).

Scheme 4. Synthesis of Phenyl-2-thiophenyl Ketone (12); Reaction of Benzoyl Chloride (14) with Thiophene (13)

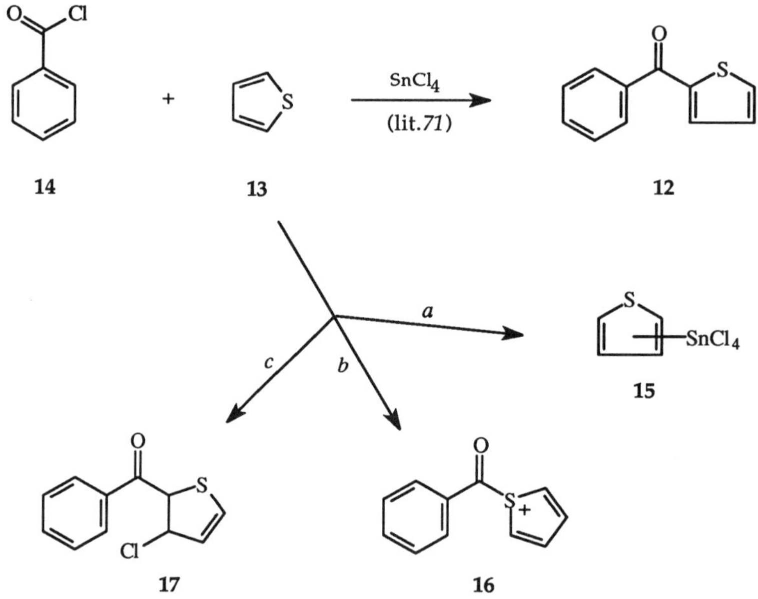

a Electrophilic Aromatic module. *b* Basic/Nucleophilic module. *c* Acidic/Electrophilic module.

The Bischler-Napieralski synthesis of 1-methyl-3,4-dihydroisoquinoline (18) from N-(2-phenylethyl)acetamide (19) in the presence of heat and acid *(72)* was not predicted by the Electrophilic Aromatic module of CAMEO; this module of CAMEO predicted instead the formation of methyl 2-acetyl-phenethylamine, 20 (Scheme 5). In a related case, CAMEO predicted (when the Electrophilic Aromatic module was chosen) that 4-anilino-butan-2-one (21) would undergo intramolecular cyclization to form 4-hydroxy-4-methyl-tetrahydroquinoline, 23 (Scheme 6). Apparently, CAMEO correctly perceives the eneamine character of 21 as necessary for this reaction to occur. It was noted in this case that ring formation, as predicted by CAMEO, depended upon the presence of mineral acid, but did not occur if a Lewis acid (e.g., stannic chloride) was selected instead as the reagent. It is known, however, that intramolecular cyclization of 21 does occur in the presence of a Lewis acid and yields 4-methyl-quinoline, 22, as product *(74)*.

LHASA Assessment. LHASA generates syntheses for user-specified target organic molecules by working backwards (retrosynthetically) from the given target to precursors which are in turn treated as new targets. The user draws a target molecule and suggests a strategy for the retrosynthetic analysis. LHASA first recognizes atoms and bond types and then perceives functional groups and various ring structures.

Scheme 5. Synthesis of 1-Methyl-3,4-dihydroisoquinoline (18) from N-(2-Phenylethyl)acetamide (19)

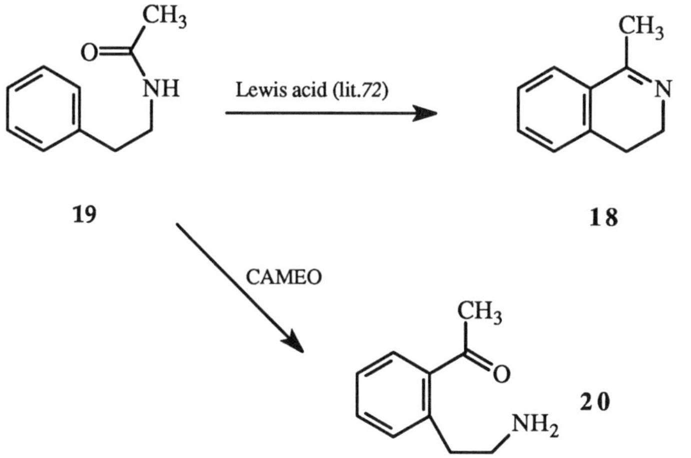

Scheme 6. Synthesis of 22 from 4-Anilino-butan-2-one (21)

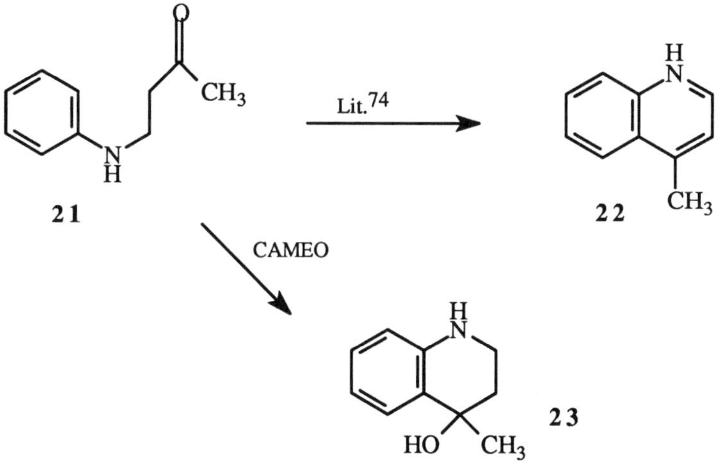

LHASA also looks for unstable functionality, bridged and fused rings, and strained substructures. Stereochemistry is assigned to the target molecule. The program then searches its knowledge base of transforms for those that satisfy the strategy selected, decides whether each transform is appropriate for the target, and displays the results. The chemist selects a precursor for further analysis and this process is repeated until precursors are found that are readily available starting materials. Each transform in the knowledge base has an initial rating that can be modified by qualifiers that consider the "favorability" of the substrate for the reaction corresponding to the current transform. More favorable reactions are given higher final ratings.

Ease of Use and User Friendliness. The LHASA program displays a Sketch Pad screen and activates "Draw Mode." A large rectangular drawing area is surrounded by three menu areas and a message window at the bottom of the screen. Cursor movement is controlled using a mouse (or cursor keys). All other screens are accessible via the Sketch Pad menus.

A chain of bonds may be drawn by repeatedly moving the cursor and then activating it by pressing the mouse button (or space bar). In Draw Mode, activating the cursor creates a carbon atom node. All atoms except hydrogen atoms are initially drawn as carbon atoms; heteroatoms are created by activating the required atom symbol in a menu of atom symbols, and then activating one or more carbon atoms causing them to be replaced by the heteroatom. The menu of atom symbols includes C, O, H, N, F, Cl, Br, I, P, S, B, Si, and Other.

Another menu area contains icons representing "templates" that allow complete cyclic and other structural elements to be entered rapidly and easily. To add a ring, the following must be activated in turn: (1) the required ring icon, (2) the first atom position of one bond in the ring, (3) the second atom position of one bond in the ring, and (4) any point in the Sketch Pad on the side of the bond where the ring is to be drawn. If steps 2 and 3 identify atom positions in an existing structure, the new ring will be automatically fused with the existing structure and the ring adjusted to avoid valence violations if possible. Two activations are required to add NO_2, COOH and SO_2 groups; these groups must always be attached to existing atoms.

The menu screens of LHASA were found to be well organized, and facilitated use of the program, particularly structure entry. Using the menu screens to identify syntheses, however, was often confusing. In addition, the HELP function generally did not provide sufficient information.

Synthesis Predictions. LHASA suggested that carbaryl could be synthesized by reaction of N-methylcarbamic acid with either 1- or 2-chloronaphthalene in the presence of strong base. The literature reference that LHASA provided and associated with this route indicates that the reaction proceeds through a benzyne intermediate. LHASA failed to suggest the commercial route *(68)* that proceeds via 1-naphthol and methyl isocyanate (Scheme 1).

LHASA predicted that 1-methyl-3,4-dihydroisoquinoline (**18**) could be synthesized from methyl phenyl N-(2-chloroethyl)imine via a Friedel-Crafts alkylation in the presence of a Lewis acid and heat. This prediction was based on a literature reference identified by LHASA. LHASA also predicted the Bischler-Napieralski synthesis of **18** from N-(2-phenylethyl)acetamide (**19**) in the presence of heat and acid (Scheme 5).

Two synthetic routes predicted by LHASA for the synthesis of quinoline (**24**) are shown in Scheme 7. LHASA's first synthetic route was the condensation of 2-aminobenzaldehyde (**25**) with acetaldehyde (**26**) to yield quinoline and water. When this transform was removed from consideration, LHASA suggested nucleophilic addition of aniline (**27**) to acrolein (**28**) to yield quinoline and water.

SYNGEN Assessment. Using a target molecule as input, SYNGEN works retrosynthetically and provides synthetic routes as output. To identify potential synthetic routes, SYNGEN performs sequential simplifications of the target structure. The first operation generates convergent assemblies of bond sets (components of the molecular skeleton), and the second operation develops abstracted functionality requirements for sequential construction from actual available starting materials. Target molecules may have up to 32 atoms or bonds. Bondsets, starting materials, intermediates, and reactions may be accessed separately for examination. Any entry in any of these categories may be individually retained or deleted to focus and prune the synthetic strategy. The number of proposed routes also may be reduced by selecting

Scheme 7. Retrosyntheses of Quinoline (24) as Predicted by LHASA

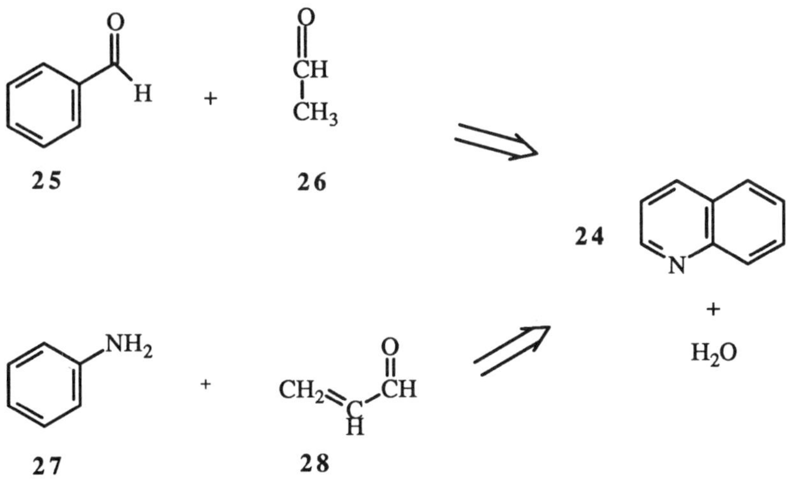

broad options such as limits on the cost of starting materials or constraining refunctionalizations of intermediates. While numerous routes are possible, SYNGEN by design limits the synthetic routes generated to the shortest and most efficient routes by adopting certain constraints: the target molecular skeleton is cut into no more than four parts, so that there are never more than 4 starting materials, all starting materials must exist in a catalog based on the Aldrich catalog (currently there are 5,926 Aldrich compounds for SYNGEN to use), and each cut severs no more than two skeletal bonds. Within these constraints, all theoretically possible routes are generated and presented for the user's evaluation. Results typically are generated within five minutes and stored for inspection at a later time. SYNGEN is not interactive and therefore independent of user bias and preconceptions.

Ease of Use and User Friendliness. From the SYNGEN main menu, the first option runs the SYNGEN molecule input and drawing interface. This is a graphic interface that allows the user to draw, name, save, and retrieve target molecular structures, and to submit structures for background processing as batch jobs by SYNGEN's reaction generator. The second option allows the user to monitor the progress of batch jobs received by the reaction generator. Each job in progress is represented by one line in a table. If all jobs are complete, no table is shown. Batch job processing appeared to start immediately upon submission, and to take no longer than 5 minutes for simple structures. Selecting the third option runs the SYNOUT interface that displays output from the reaction generator. The results were ready for review typically in 10 minutes or less. The user may select a target structure from a list of available structures, and may then review separate graphics screens displaying:
- the target structure and summary statistics,
- bondsets,
- starting materials,
- intermediate structures, and
- reactions.

The screen is divided into a drawing area and three menu areas. A cross-wire cursor can be used in the drawing area to input bond lines, and in the menu areas to select menu items. Alternatively, menu items may be selected by keystrokes which are displayed in the menu. One of the three menu areas contains a useful list of templates allowing complete structural elements to be entered in one action. For example, the template for Aromatic rings allows a ring to be drawn by pressing "T", "A", and a number key to indicate the number of carbon atoms in the ring. Some menu items were intuitive and worked well consistently. These include Center, Reduce, Enlarge, Erase Last, Clean, Kill Image, and Refresh. Others including Relocate, Bond Change, Atom Change, Hold, Fix Struct., Make User, Get User, and Read were less obvious. Experimentation often obtained the desired results. Limited context-sensitive Help was available.

Some apparent limitations of the SYNGEN interface include the following: bonds added in error could be removed using the "Erase Last" command, but there is no "undo" command that would allow the last deleted bond to be reinstated; the system does not seem to encourage editing of existing structures to give new structures; "Delete Bond" allows bonds to be deleted, but there is no "Create Bond" command to merge two independent structures into one.

Synthesis Prediction. The most straightforward and cost-effective method proposed by SYNGEN for the synthesis of carbaryl (**3**) (out of 464 suggested routes utilizing 111 different starting materials) was the nucleophilic substitution reaction of N-methylcarbamoyl chloride with 1-naphthol. The reaction of 1-naphthol with methyl isocyanate was not predicted even though methyl isocyanate appears in the starting materials database of SYNGEN.

SYNGEN proposed the reaction of ethyl chloroacetate with 4-vinylpyridine (**5**) as the optimal route to 4-(2-carbethoxycyclopropyl)pyridine (**6**). The reaction mechanism, which has been documented *(69)*, was characterized by SYNGEN as alkylation via hydrogen substitution followed by conjugate addition. For this compound, 11,031 routes utilizing 239 different starting materials were suggested, indicative of the breadth and scope of routes possible with SYNGEN.

Out of a total of 2,267 routes utilizing 78 different starting materials, SYNGEN's first choice for the synthesis of 5,8,9,10-tetrahydro-1,4-naphthoquinone was the Diels-Alder reaction using 1,3-butadiene and 1,4-benzoquinone. This pathway is reported as quantitative when run in benzene heated to 35 ^{0}C *(70)*.

The two most cost-effective pathways (out of 135 different routes) proposed by SYNGEN for the synthesis of 4-nitrodiphenylamine (**29**) were the aromatic substitution reactions of aniline (**27**) with a *para*-halonitrobenzene (**30**), or *para*-nitroaniline (**31**) with a halobenzene (**32**) (Scheme 8).

Discussion

Advancements in computer technology within the past decade have greatly facilitated scientific research and development in essentially all scientific disciplines. Improvements in computer graphics, computational power and programming have led, for example, to the development of highly sophisticated chemistry-related software programs that enhance the chemist's ability to: elucidate structures; understand reaction mechanisms; identify synthesis routes; and predict physical-chemical properties, to name a few. Chemical synthesis software programs such as CAMEO, LHASA and SYNGEN have had programmed within them chemical theory and mechanistic "rules," which are used in computing reaction pathways or, in the case of CAMEO, reaction products. This approach to identifying reaction pathways or products is similar to that of a practicing chemist, who has learned the mechanistic rules of chemical theory and applies this knowledge to plan a synthesis. Our intent in this study was to assess the usefulness of synthesis software programs for the ultimate purpose of assisting EPA chemists (and their industrial counterparts) in identifying theoretically less polluting

Scheme 8. Retrosyntheses of 4-Nitrodiphenylamine (29) as Predicted by SYNGEN.

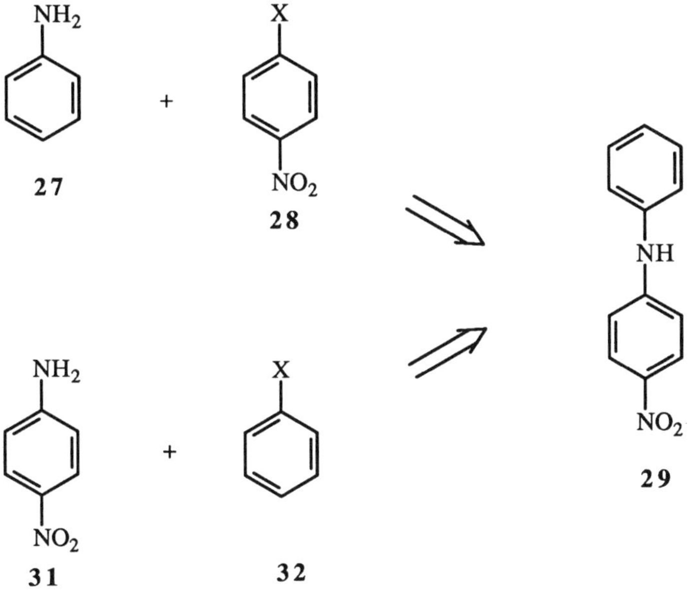

synthetic pathways as alternatives to synthesis pathways encountered, for example, in EPA's Premanufacture Notification (new chemicals) program and existing chemicals program. Although current synthesis software programs do not incorporate pollution prevention strategies in synthesis design, we felt that using such software programs should at least provide more theoretical synthesis routes than a chemist may deduce independently. By having more reaction pathways to choose from, a chemist will have an increased likelihood of identifying reaction pathways that are less polluting.

Our preliminary evaluation of CAMEO, LHASA, and SYNGEN was confined to their overall user-friendliness, their ability to predict syntheses that are similar or identical to literature (known) syntheses, and their ability to provide rational alternative syntheses that may not appear in the literature. Caution must be used when assessing the predictive quality of computer-generated syntheses by comparison to literature syntheses. For any given organic substance there are a number of theoretically plausible syntheses that could be used to prepare the substance. Even if none of these syntheses appears in the literature, one should not conclude that the computer syntheses are not valid or will not work if tried experimentally. In this study we purposely selected chemical substances whose known (reported) syntheses involve commonly encountered reactions (e.g., Diels-Alder cycloaddition, Friedel-Crafts acylation, etc.) and commonly encountered reagents. Because the software programs used in this study apply chemical theory and mechanistic rules that are representative of the known syntheses of the selected chemicals, we felt that comparing the computer-generated syntheses to literature syntheses of these chemicals would be useful in assessing how well these software programs apply the theory and rules that have been incorporated into them.

CAMEO. CAMEO operates in the synthetic direction and in theory could be used to confirm the reasonableness of potential synthetic pathways under review and to indentify potential by-products and side-reactions that may also be generated.

Ease of Use and User Friendliness. CAMEO was found to be easy to use. CAMEO's menu screens are well designed and easy to follow, and greatly facilitate use and operation of the program, particularly structure entry. To explore fully all potential reactions and the products of these reactions, however, the user must separately evaluate a given set of starting materials and reaction conditions under several if not all of the available reaction modules (i.e., Carbenoid, Radical, Heterocyclic, Basic/Nucleophilic, Acidic/Electrophilic, Electrophilic Aromatic, Oxidative/Reductive, and Pericyclic). Thus, the user can enter reactants and reaction conditions, and, depending upon which module is selected, CAMEO may predict different results. For example, CAMEO correctly predicted carbaryl as the product from the reaction of methyl isocyanate with 1-naphthol only if the Acidic/Electrophilic mechanistic module was selected; no product was predicted when the Basic/Nucleophilic module was selected. 1-Naphthol is clearly the nucleophile in this reaction, and it seems that CAMEO should have recognized it as such.

The need for the user to select a module is problematic for two reasons. First, reactants will always react consistently to form a product or products so long as reaction conditions are kept the same. Depending upon which module is selected, CAMEO will give different products even if reactants and reactions conditions are kept the same. Second, the type of mechanism involved in the reaction of two or more chemicals under given conditions may not always be known (or at least may not be obvious) to one using CAMEO. For the CAMEO program to be more useful, especially for the applications envisioned by EPA, it should have the ability to assess the reactive nature of reactants under certain conditions, and therefore would not need to be "told" what the reaction "type" is. The necessity of selecting the appropriate or, in the case of carbaryl, the inappropriate module(s) to evaluate the reaction potential of reactants appears to be a significant limitation of CAMEO.

Synthesis Prediction. CAMEO correctly predicted the synthesis of **6** from the reaction of **4** with **5** (Scheme 2). That CAMEO predicted this reaction only when ultraviolet light was selected to initiate the reaction (when it is known that this reaction proceeds when heated (*69*)) should not be interpreted as an error of CAMEO. It is well known that diazo carbonyl compounds (*e.g.*, **4**) react under photolytic conditions (*75*). It would be interesting to determine experimentally the actual products formed from the reaction of **4** with **5** using ultraviolet light, to observe if the reaction products predicted by CAMEO (compounds **6** and **7**) are in fact formed.

The yield of product **10** from the Diels-Alder cycloaddition reaction of 1,3-cyclohexadiene (**8**) with maleic anhydride (**9**, Scheme 3) is reported to be quantitative when run in warm benzene *(70)*. The fact that CAMEO predicted **10** as a minor product and **11** as the major product from the reaction of **8** with **9** suggests that CAMEO has limited applicability with Diels-Alder-type cycloaddition reactions. Furthermore, CAMEO failed to predict the correct product (**12**) via Friedel-Crafts acylation of thiophene (**13**) with **14** (Scheme 4). Thiophene is highly reactive under these conditions, and CAMEO was expected to predict **12**, which is the known reaction product of this reaction (*71*). The CAMEO-predicted products **15-17** also demonstrate how module selection affects output, even if the reaction conditions are the same. These results demonstrate that CAMEO may not identify well-characterized reaction pathways and their predominant reaction products.

At first glance it would appear that CAMEO did not predict the correct product from the treatment of **19** with heat and a Lewis acid (the known synthesis of 1-methyl-3,4-dihydroisoquinoline, **18**, ref. *72*) as shown in Scheme 5. 2-Acetyl-phenethylamine, **20**, (the product predicted by CAMEO), however, is a known immediate precursor of **18** *(76)*. Compound **20** can also be viewed as the hydrolysis product of **18**, and it may be that CAMEO predicted formation of **18** as an intermediate that undergoes hydrolysis to yield **20** (it seems reasonable to assume that **18** is unstable in water and will hydrolyze quickly to **20**). Prediction of **20** as the product from the reaction of **19** with a Lewis acid and heat should not be viewed as a fault of CAMEO.

The treatment of 4-anilinobutan-2-one (**21**, Scheme 6) with acid has been shown to give a variety of products, depending upon the types of solvents and acids used *(74)*. When ethanol is the solvent and $FeCl_3$ (a Lewis acid) is used, 4-methylquinoline (**22**) is obtained in quantitative yield *(74)*. When ethanolic HCl is used, however, reaction products include: **22**; aniline; butan-2-one; and 4-ethoxybutan-2-one. The yield of butan-2-one is equimolar to **22**. The same reaction in dioxane gave methyl vinyl ketone instead of butan-2-one, together with the above products. Ogata, *et al.* *(74)* suggested that the synthesis of **22** from **21** proceeds via formation of **23**, which is an intermediate that undergoes sequential dehydrogenation and dehydration to yield **22**. We feel that CAMEO's prediction of **23** as the product from the reaction of **21** with acid is largely correct. However, CAMEO correctly predicted the formation of **23** only in the presence of mineral acid (*e.g.*, HCl); **23** was not predicted when a Lewis acid was chosen, which is inconsistent with the experimental results described above.

LHASA.

Ease of Use and User Friendliness. LHASA is highly interactive and requires the user to make expert judgments to determine the retrosynthetic tree(s) generated and evaluated. To operate LHASA properly, the user must possess a considerable amount of chemical knowledge and organic synthetic design skill. For example, the chemist determines which precursor to process next and selects a strategy for processing. These choices have a major impact upon outcome of the retrosynthetic analysis. The system searches a knowledge base of transforms and selects those that match both the strategy and the precursor. The system then presents a new level of suggested precursors. The user may choose to process some precursors further and ignore others.

LHASA offers the user five alternative strategies:
1. Short-range strategy
2. Topological strategy
3. Long-range strategy
4. Stereochemical strategy
5. Starting-material-oriented strategy

Strategies 1, 2, and 4 generate broad retrosynthetic trees with fewer levels, and require substantial user interaction. Strategies 3 and 5 generate narrower trees with more levels and require less user interaction. Since the user re-selects a strategy for each precursor, several different strategies may be combined in the same synthetic path. In general, the high degree of user interaction makes identifying and evaluating a broad range of alternative synthetic pathways a time-consuming, labor-intensive process.

Synthesis Predictions. In this study LHASA was found to propose reaction pathways that are either essentially identical to known (reported) syntheses or sufficiently plausible from a chemistry standpoint. LHASA suggested that carbaryl could be synthesized by reaction of N-methylcarbamic acid with either 1- or 2-chloronaphthalene in the presence of strong base. The literature reference that LHASA provided and associated with this route indicates that the reaction proceeds through a benzyne intermediate. No literature reference was found, however, that describes the synthesis of carbaryl using 2-chloronaphthalene. Surprisingly, LHASA failed to suggest the well known commercially used route *(68)* that proceeds via 1-naphthol and methyl isocyanate (Scheme 1).

LHASA predicted that 1-methyl-3,4-dihydroisoquinoline (**18**, Scheme 5) could be synthesized from methyl phenyl N-(2-chloroethyl)imine via a Friedel-Crafts alkylation in the presence of a Lewis acid and heat. Although this predicted synthesis of **18** does not appear in the literature, it appears to be a highly plausible alternative to the known syntheses of **18**. In fact, the ability of LHASA to provide plausible alternative syntheses to known syntheses is a desirable quality of the program. LHASA also predicted the synthesis of **18** from N-(2-phenylethyl)acetamide (**19**) in the presence of heat and acid (Scheme 5), which is identical to a known synthesis for this compound *(72)*.

Two synthetic routes predicted by LHASA for the synthesis of quinoline (**24**) are shown in Scheme 7. LHASA's first synthetic route was to have 2-aminobenzaldehyde (**25**) condense with acetaldehyde (**26**) to yield quinoline and water, which is essentially equivalent to the known synthesis of quinoline from the reaction of aniline with acetaldehyde and formaldehyde *(77)*. When this transform was removed from consideration, LHASA suggested nucleophilic addition of aniline (**27**) to acrolein (**28**) to yield quinoline and water, which is identical to a known synthesis of quinoline *(73)*.

SYNGEN.

Ease of Use and User Friendliness. The SYNGEN system limits target organic molecules to no more than 32 atoms or bonds, and to syntheses from not more than four starting structures. SYNGEN synthetic routes are limited to those which in theory generate no by-products; SYNGEN does not by itself facilitate the exploration of by-products arising from different routes of synthesis. There appears to be no way to indicate stereochemistry. The package could benefit by incorporating additional user-friendly features such as file directories, cut and paste between screens, automatic save, and an "undo" command. During review, some messages were displayed and removed too rapidly to be read. It was also noted that SYNGEN allows several distinct structures to be given the same name.

SYNGEN is limited to target molecules of up to 32 atoms or bonds. In some instances this could be a significant limitation for EPA's intended applications. We did not explore these limits. It is not clear whether the user needs to count bonds or atoms or both, whether the system warns the user when the limit is reached, and what kinds of errors might occur if the limit is exceeded.

Synthesis Prediction. The most straightforward and cost-effective method proposed by SYNGEN for the synthesis of carbaryl (**3**) (out of 464 suggested routes utilizing 111 different starting materials) was the nucleophilic substitution reaction of N-methylcarbamoyl chloride with 1-naphthol. Surprisingly, the known reaction of 1-naphthol with methyl isocyanate to yield carbaryl was not predicted even though methyl isocyanate appears in the starting materials database of SYNGEN.

SYNGEN proposed the reaction of ethyl chloroacetate with 4-vinylpyridine (**5**) as the optimal route to 4-(2-carbethoxycyclopropyl)pyridine (**6**). The reaction mechanism, which has been documented *(69)*, was characterized by SYNGEN as alkylation via hydrogen substitution followed by conjugate addition. For this compound, 11,031 routes utilizing 239 different starting materials were suggested, indicative of the breadth and scope of routes available through SYNGEN.

Out of a total of 2,267 routes utilizing 78 different starting materials, SYNGEN's first choice for the synthesis of 5,8,9,10-tetrahydro-1,4-naphthoquinone was the Diels-Alder reaction using 1,3-butadiene and 1,4-benzoquinone. This reaction is reported as quantitative when run in benzene heated to 35 °C *(70)*.

4-Nitrodiphenylamine (**29**, Scheme 8) is currently produced industrially from the reaction of 4-chloronitrobenzene and aniline. This reaction, in which aniline displaces the chlorine atom of 4-chloronitrobenzene, results in the generation of large quantities of hydrogen chloride and halogenated waste. The two most cost-effective pathways (out of 135 different routes) proposed by SYNGEN for the synthesis of 4-nitrodiphenylamine (**29**, Scheme 8) were the known aromatic substitution reactions of aniline (**27**) with a *para*-halonitrobenzene (**30**), or *para*-nitroaniline (**31**) with a halobenzene (**32**). The method recently announced by Monsanto *(78)* of generating this product directly from aniline and nitrobenzene, which eliminates these problems, was not predicted by SYNGEN.

Conclusion

CAMEO, LHASA, and SYNGEN each possess attributes that are potentially useful in supporting pollution prevention initiatives at EPA. No one program, however, addresses all the chemical synthesis issues that arise during review of new and existing chemicals. Also, the reliability and completeness of the pathways suggested or products proposed is variable but is expected to improve as the authors of these programs continue to enhance the software.

CAMEO can help identify side reactions and by-products that may be generated in the synthetic route described for example in a Premanufacture Notification or by a synthetic pathway proposed by a retrosynthetic CAOS program. These materials often may be of concern to Agency scientists. In its present form, however, with each reaction module having to be selected by the user, CAMEO cannot automatically screen for potential reactivity of reactants under review. Further, within a module, reaction conditions such as temperature, solvent, and choice of catalyst cannot be varied automatically to look for reactivity, again because these options must be selected by the user. A useful enhancement would be to have the option of directing CAMEO to select all modules, to vary reaction conditions, and to display the set of all unique products found and their route(s) of synthesis. CAMEO also can be used to help confirm synthetic pathways proposed by retrosynthetic CAOS programs and thus aid Agency chemists to select chemically valid alternative synthetic pathways.

LHASA allows for and requires significant user interaction to explore and develop reaction pathways. LHASA provides literature references to substantiate each proposed retrosynthetic step. Synthetic routes proposed by LHASA in general were judged to be chemically plausible. LHASA was found to be cumbersome, however, in exploring alternative synthetic pathways: when trying to develop a large set of possible synthetic routes for a given compound, LHASA was labor-intensive and time-consuming. A useful enhancement to LHASA would be the ability to propose and display a user-defined number of initial pathways for further analysis.

SYNGEN quickly generates in a single operation numerous alternative pathways to achieve the synthesis of most target compounds with little input from the user. SYNGEN is a useful tool for quickly identifying pathways for further analysis. Using the default parameters however, many pathways developed by SYNGEN would not be economically or chemically realistic. Cost data provide a useful parameter for ranking of proposed routes, but may be out of date and not reflective of actual costs for industrial quantities of starting materials. It appears possible that other parameters such as acute or chronic toxicity, ecotoxicity, etc. could be incorporated into the ranking scheme as well. A practical enhancement to SYNGEN (and perhaps to CAMEO and LHASA) that would help focus and limit the number of potential pathways would be to permit the user to enter a molecule as a precursor and direct SYNGEN to find syntheses that used this compound. Conversely, the ability of the user to remove toxic or hazardous precursors from consideration also would be a helpful added feature.

Acknowledgments

The authors thank Drs. Roger L. Garrett and Joseph Breen for their support and encouragement. This work was supported under contract number 68-D0-0020 for the U.S. Environmental Protection Agency to the Dynamac Corporation, Rockville, MD. The authors thank Drs James B. Hendrickson (Brandeis University) and Alan Long (Harvard University) for contributing the versions of SYNGEN and LHASA, respectively, used in this study.

Disclaimer

The research described in this paper has been reviewed by the Office of Pollution Prevention and Toxics, U.S. Environmental Protection Agency, and approved for publication. Approval does not signify that the contents necessarily reflect the views and policies of the Agency nor does mention of commercial products constitute endorsement or recommendation for use.

Literature Cited

1. Pollution Prevention Act of 1990. 42 U.S.C. §§13101-13109, **1990**.
2. Browner, C.M. *EPA Journal* **1993**, *19*, pp 6-8.
3. Draths, K.M.; Ward, T.L.; Frost, J.W. *J. Am. Chem. Soc.* **1992**, *114*, pp 9725-9726.
4. Illman, D.L. *Chem. Eng. News* **Sept 6, 1993,** pp 26-30.
5. Agarwal, K.K.; Larsen, D.L.; Gelernter, H.L. *Comput. Chem.* **1978**, *2*, pp 75-84.
6. Barone, R.; Chanon, M. In *Computer Aids to Chemistry*; Vernin, G.; Chanon, M., Eds.; John Wiley & Sons: New York, NY, 1986; pp 19-102.
7. Barone, R.; Arbelot, M.; Chanon, M. *Tetrahedron Comput. Methodol.* **1988**, *1*, pp 3-14.
8. Bauer, J. *Tetrahedron Comput. Methodol.* **1989**, *2*, pp 269-280.
9. Bauer, J.; Fontain, E.; Forstmeyer, D.; Ugi, I. *Tetrahedron Comput. Methodol.* **1988**, *1*, pp 129-132.
10. Bauer, J.; Herges, R.; Fontain, E.; Ugi, I. *Chimia* **1985**, *39*, pp 43-53.
11. Baumer, L.; Sala, G.; Sello, G. *Anal. Chim. Acta* **1990**, *235*, pp 209-214.
12. Bertrand, M.P.; Monti, H.; Barone, R. *J. Chem. Ed.* **1986**, *63*, p 624.
13. Blurock, E.S. *Tetrahedron Comput. Methodol.* **1989**, *2*, pp 207-222.
14. Blurock, E.S. *J. Chem. Inf. Comput. Sci.* **1990**, *30*, pp 505-510.
15. Brandeis University, Department of Chemistry. *The SYNGEN program for synthesis generation: Demonstration of input/output*. Brandeis University: Waltham, MA, 1989.
16. Bures, M.G.; Jorgensen, W.L. *J. Org. Chem.* **1988**, *53*, pp 2504-2520.
17. Burden, F.R. *Tetrahedron Comput. Methodol.* **1988**, *1*, pp 169-176.
18. Carlsen, Per H.J.; Edvardsen, O. *Tetrahedron Comput. Methodol.* **1989**, *2*, pp 305-308.
19. Choplin, F.; Bonnet, P.; Zimmer, M.Z.; Kaufmann, G. *Nouv. J. Chim.* **1979**, *3*, pp 223-230.
20. Corey, E.J.; Johnson, A.P.; Long, A.K. *J. Org. Chem.* **1980**, *45*, pp 2051-2057.
21. Corey, E.J.; Long, A.K.; Rubenstein, S.D. *Science* **1985**, *228*, pp 408-418.
22. Elrod, D.W.; Maggiora, G.M.; Trenary, R.G.*J. Chem. Inf. Comput. Sci.* **1990**, *30*, pp 477-484.
23. Figueras, J. *SynTree: A program for exploring organic synthesis*. Trinity Software: Campton, NH, **1992.**

24. Funatsu, K.; Del Carpo, D.A.; Sasaki, S. *Tetrahedron Comput. Methodol.* **1988**, *1*, pp 39-51.
25. Funatsu, K.; Endo, T.; Kotera, N.; Sasaki. S. *Tetrahedron Comput. Methodol.* **1988**, *1*, pp 53-69.
26. Funatsu, K.; Sasaki, S. *Tetrahedron Comput. Methodol.* **1988**, *1*, pp 27-37.
27. Gasteiger, J.; Jochum, C. *Top. Curr. Chem.* **1978**, *74*, pp 93-126.
28. Gasteiger, J.; Ihlenfeldt, W.D.; Rose, P.; Wanke, R. *Anal. Chim. Acta* **1990**, *235*, pp 65-75.
29. Gasteiger, J.; Marsili, M.; Hutchings, M.G.; Saller, H.; Low, P.; Rose, P.; Rafeiner, K. *J. Chem. Inf. Comput. Sci.* **1990**, *30*, pp 467-476.
30. Gelernter, H.L.; Sanders, A.F.; Larssen, D.L.; Agarwal, K.K.; Boivie, R.H.; Spritzzer, G.A.; Searleman, J.E. *Science* **1977**, *197(4308)*, pp 1041-1049.
31. Gelernter, H.; Rose, J.R.; Chen. C. *J. Chem. Inf. Comput. Sci.* **1990**, *30*, pp 492-504.
32. Hendrickson, J.B. *Acc. Chem. Res.* **1986**, *19*, pp 274-281.
33. Hendrickson, J.B. *Angew. Chemie, Int. Ed. Engl.* **1990**, *29*, pp 1286-1295.
34. Hendrickson, J.B.; Braun-Keller, E. *J. Comput. Chem.* **1980**, *1*, pp 323-333.
35. Hendrickson, J.B.; Miller, T.M. *J. Chem. Inf. Comput. Sci.* **1990**, *30*, pp 403-408.
36. Hendrickson, J.B.; Miller, T.M. *J. Am. Chem. Soc.* **1991**, *113*, pp 902-910.
37. Hendrickson, J.B.; Miller, T.M. *J. Org. Chem.* **1992**, *57*, pp 988-994.
38. Hendrickson, J.B.; Toczko, A.G. *J. Chem. Inf. Comput. Sci.* **1989**, *29*, pp 137-145.
39. Herges, R. *Tetrahedron Comput. Methodol.* **1988**, *1*, pp 15-25.
40. Herges, R. *J. Chem. Inf. Comput. Sci.* **1990**, *30*, pp 377-383.
41. Hippe, Z. 1981. *Anal. Chim. Acta* **1981**, *133*, pp 677-683.
42. Jiang, K.; Zheng, J.; Higgins, S.B.; Watterson, D.M.; Craig, T.A.; Lukas, T.J.; Van Eldik, L.J. *Comput. Appl. Biosci.* **1990**, *6*, pp 205-212.
43. Johnson, A.P. *Chem. Brit.* **1985**, *21*, pp 59-67.
44. Jorgensen, W.L.; Laird, E.R.; Gushurst, A.J.; Fleicher, J.M.; Gothe, S.A.; Helson, H.E.; Paderes, G.D.; Sinclair, S. *Pure Appl. Chem.* **1990**, *62*, pp 1921-1932.
45. Kudo, Y.; Suzuki, T. *Bull. Yamagata Univ. (Engl.)* **1989**, *20*, pp 207-219.
46. Laird, E.R.; Jorgensen, W.L. *J. Chem. Inf. Comput. Sci.* **1990**, *30*, pp 458-466.
47. Mager, P.P. *Med. Res. Rev.* **1991**, *11*, pp 375-402.
48. Marsili, M. 1990. In *Computer Chemistry*; CRC Press, Inc: Boca Raton, FL, 1990; pp 141-193.
49. *Brainstorm with the best minds in chemistry. REACCS.* Molecular Design Ltd:San Leandro, CA, **1990**.
50. Moll, R. *Anal. Chim. Acta* **1990**, *235*, pp 189-193.
51. Nevalainen, V.; Pohjala, E.; Malkonen, P.; Hukkanen, H. *Acta Chem. Scand.* **1990**, *44*, pp 591-602.
52. Olsson, T. *Acta Pharm. Suec.* **1986**, *23*, pp 386-402.
53. Paderes, G.D.; Jorgensen, W.L. *J. Org. Chem.* **1989**, *54*, pp 2058-2085.
54. Parlow, A.; Weiske, C.; Gasteiger, J. *J. Chem. Inf. Comput. Sci.* **1990**, *30*, pp 400-402.
55. Peishoff, C.E.; Jorgensen, W.L. *J. Org. Chem.* **1983**, *48*, pp 1970-1979.
56. Pollet, P. *J. Chem. Ed.* **1986**, *63*, pp 624-625.
57. Presnell, S.R.; Benner, S.A. *Nucl. Acids Res.* **1988**, *16*, pp 1693-1702.
58. Rose, P.; Gasteiger, J. *J. Anal. Chim. Acta* **1990**, *235*, pp 163-168.
59. Salatin, T.D.; Jorgensen, W.L. *J. Org. Chem.* **1980**, *45*, pp 2043-2051.

60. Sanderson, D.M.; Earnshaw, C.G. *Hum. Exper. Toxicol.* **1991**, *10*, pp 261-273.
61. Brandt, J.; Friedrich, J.; Gasteiger, J.; Jochum, C.; Schubert, W.; Ugi, I. In *Computer-Assisted Organic Synthesis;* Wipke, T.D.; Dyott, T.M., Eds.; American Chemical Society Symposium Series 61; American Chemical Society: Washington, DC, 1977; pp 33-59.
62. Ugi, I.K.; Fontain, E.; Bauer, J. *Anal. Chim. Acta* **1990**, *235*, pp 155-161.
63. Tunghwa, W.; Burnstein, I.; Corbett, M.; Ehrlich, S.; Evens, M.; Gough, A.; and Johnson, P. In American Chemical Society Symposium Series No. 306; American Chemical Society: Washington, DC, 1986; pp 244-257.
64. Warr, W.A.; Suhr, C.*Chemical Information Management;* VCH Publishers, Inc.: New York, NY, 1992; p 127.
65. Weiner, M.P.; Scheraga, H.A. *Comput. Appl. Biosci.* **1989**, *5*, pp 191-198.
66. Wipke, W.T.; Braun, H.; Smith, G.; Choplin, F.; Sieber, W. In *Computer-Assisted Organic Synthesis;* Wipke, T.D.; Dyott, T.M., Eds.; American Chemical Society Symposium Series 61; American Chemical Society: Washington, DC, 1977: pp 97-127.
67. Wipke, W.T.; Ouchi, G.I.; Krishnan, S. *Artif. Intell.* **1978**, *11*, pp 173-193.
68. Thadeo, P.T.; Movery, D.F. *J. Chem. Ed.* **1984**, *61*, p 742.
69. Gray, A.P.; Kraus, H. *J. Org. Chem.* **1966**, *31*, pp 399-405.
70. *Organic Chemistry;* Morrison, R.T.; Boyd, R.N., Eds.; 3rd Edition; Allyn and Bacon, Inc.: Boston, MA, 1973; p 877.
71. ibid., p 1008.
72. ibid., p 1021.
73. ibid., pp 1019-1020.
74. Ogata, Y.; Kawasaki, A.; Suyama, S. *J. Chem. Soc. B,* **1969**, *7*, pp 805-810.
75. *Advanced Organic Chemistry;* March, J., Ed.; 3rd Ed.; John Wiley & Sons, Inc.: New York, NY, 1985; pp 974-975.
76. Pictet, K.*Chem. Ber.,* **1909**, *42*, p 1977.
77. United States Patent 3,020,281; **1962** to Reilly Tar & Chemical.
78. Stern, M.K.; Bashkin, J.K. U.S. Patent 5117063A 920526; **1992** to Monsanto.

RECEIVED August 4, 1994

INDEXES

Author Index

Anastas, Paul T., 2,155,166
Blackert, Joseph F., 98
Cavanaugh, Margaret A., 23
Chapman, Orville L., 114
DeVito, Stephen C., 166
Draths, Karen M., 32
Epling, Gary A., 64
Farris, Carol A., 155
Frost, John W., 32
Gargulak, Jerry D., 46
Gladfelter, Wayne L., 46
Hancock, Kenneth G., 23
Kirihara, Masayuki, 76
Kraus, George A., 76
Manzer, Leo E., 144
McCarthy, Bridget A., 84
McGhee, William D., 122
Nies, J. Dirk, 166
Podall, Harold E., 155
Riley, Dennis, 122
Sadeghipour, Mitra, 98
Snider, Barry B., 84
Stern, Michael K., 133
Tanko, James M., 98
Waldman, Thomas, 122
Wang, Qingxi, 64
Wu, Yusheng, 76

Affiliation Index

Brandeis University, 84
DuPont, 144
Dynamac Corporation, 166
Iowa State University, 76
Michigan State University, 32
Monsanto Company, 122,133
National Science Foundation, 23
U.S. Environmental Protection Agency, 2,155,166
University of California–Los Angeles, 114
University of Connecticut, 64
University of Minnesota, 46
Virginia Polytechnic Institute and State University, 98

Subject Index

A

Academic research, role in benign by design chemistry, 11
2-Acylhydroquinones, synthesis, 78–79
Additions and cyclizations, $Mn(OAc)_3$-based oxidative free-radical, 85–86
Adipic acid
  applications, 32
  synthesis from D-glucose, 33–44
  synthesis from toxic materials, 32–34f
  waste resourcing, 148
Adiponitrile, waste resourcing, 146,148
Alkylaromatics, free-radical bromination, 103–104
Alternative synthetic design, computer-assisted, for pollution prevention, 166–183
Amines
  aromatic, See Aromatic amines
  to generate urethanes and isocyanates, 122–131
4-Aminodiphenylamine, synthesis, 134–139

Anodic oxidation, use for organic synthesis, 86
Aromatic amines
  halide-free routes for synthesis, 133–142
  intermediates in metal-catalyzed carbonylation of nitroaromatics to aryl carbamates, 46
Asymmetric catalysis, applications, 149–150
Asymmetric hydrocyanation, synthesis of precursor to Naproxen, 149
Atom economy, concept, 76

B

Benign by design chemistry
  alternative catalysis, 13
  alternative feedstocks, 12–16
  alternative processes, 16
  approaches, 11–15f
  catalysis, 13
  computer design of synthetic methodologies, 12
  economic viability, 10
  future, 19–20
  governmental initiatives, 16–18
  industrial initiatives, 13,16
  materials of consequence, 11
  public–private partnerships, 16–18
  role of academic research, 11
  solvent alternatives, 12
  synthetic efficiency, 10
Benign synthetic design tool development, initiatives for benign by design chemistry, 17
Benzene
  activation by chlorination, 133,134f
  for adipic acid synthesis, 32
  in chemical industry, 33
  in styrene manufacture, 114–115
Benzophenone, substituted, synthesis via photochemical alternative to Friedel–Crafts reaction, 78–79
Benzyl protecting group, cleavage, 71–73

Bicyclo[3.2.1]octanone, synthesis, 86
$endo$-Bicyclo[2.2.2]oct-5-ene-2,3-dicarboxylic anhydride, synthetic prediction using CAMEO, 171
Biocatalysis, microbial, 39–44
Bromination
  competitive, 106–108
  free-radical, 103–104,110–111
  Ziegler, 108–110

C

CAMEO
  assessment for computer-assisted alternative synthetic design, 167–173
  ease of use and user friendliness, 178
  synthetic prediction, 178–179
  translation of chemical knowledge into synthetic proposals, 169
Carbaryl, synthetic prediction using CAMEO, 169
4-(2-Carbethoxycyclopropyl)pyridine, synthetic prediction using CAMEO, 170–171
Carbon dioxide
  supercritical, as medium for free-radical reactions, 98–112
  to generate urethanes and isocyanates, 122–131
Carbonylation of nitroaromatics, mechanistic study of catalytic process, 46–62
Catalysis
  asymmetric, 149–150
  benign by design chemistry, 13
  environmentally safer product production, 151–154
  hazardous and toxic material management, 150–151
  role in waste conversion to salable products, 144
  waste minimization and resourcing, 145–149

INDEX

Catalytic process for carbonylation of nitroaromatics
  C–N bond-forming step, 56–57
  experimental procedure, 47–50
  impact on goal of replacing phosgene, 59,62
  isomer formation, 50
  mechanism of overall catalytic cycle, 59,61f
  product-forming step, 57–59,60f
  reactivity of Ru complex, 51–54
Chemical industry, economics, 23–24
Chemical manufacturing
  changes, 78
  costs, 7
    industries with most toxic substance releases, 7,8f
  yield, 67
Chemical reoxidation, Mn(II) to Mn(III), 93
Chemical substances, new, pollution potential, 156
Chemical syntheses, source of pollution, 167
Chemist, importance in pollution prevention, 9–10
Chemistry
  environmentally safer processes, 144–154
  problems created, 23
Chiral phosphinite ligands, use for hydrocyanation, 150
Chlorine atoms, role in ozone depletion, 99
Chlorobenzenes, nitration, 133,134f
Chlorofluorocarbons, potential substitutes, 151–154
Competitive bromination, ethylene and toluene, 106–108t
Computer-assisted alternative synthetic design for pollution prevention, 166–183
  CAMEO assessment, 169–173,178–179
  goal, 168
  LHASA assessment, 172–175,179–180
  program evaluation procedure, 169
  software selection criteria, 168–169
  SYNGEN assessment, 174–177,180–181
Computer design of synthetic methodologies, benign by design chemistry, 12
Control and treatment technologies of pollution, costs, 3
Curriculum development, initiatives for benign by design chemistry, 17
Cyclizations, Mn(OAc)$_3$-based oxidative free-radical, 85–86
Cyclopropylbenzene, free-radical bromination, 110–111

D

Department of Energy, initiatives for benign by design chemistry, 18
Diazepam, synthesis via photochemical alternative to Friedel–Crafts reaction, 77,80–81
Dibasic acids, resourcing, 148
Dibasic esters, resourcing, 148
Dilution as solution to pollution, 23
Dithiane protecting group, cleavage, 67–70
Dithio acetals and ketals, use in organic synthesis, 66
Doxepin, 76
Dye-promoted photocleavage of dithio compounds to aldehydes and ketones, mechanism, 69

E

Economic pressures for environmental preservation, 24
Economic viability, role in benign by design chemistry, 10
Efficiency of synthetic methodology, role in benign by design chemistry, 10
Electrochemical oxidative free-radical cyclizations, Mn(III)-mediated, 84–94
End-of-pipe approach, regulation of hazardous chemicals, 166–167

Environmental chemistry
  research at National Science Foundation, 28–29
  research challenges, 27–29
  research partnership between National Science Foundation and Environmental Protection Agency, 29
Environmental preservation, economic and social pressures, 24
Environmental Protection Agency, *See* U.S. Environmental Protection Agency
Environmental questions, benign by design chemistry, 110
Environmental technology, initiatives for benign by design chemistry, 18
Environmentally benign chemical synthesis and processing, National Science Foundation–Council for Chemical Research program, 24–27
Environmentally safer processes
  hazardous and toxic material management, 150–151
  high-yield–low-waste processes, 145
  product production, 151–154
  waste minimization and resourcing, 145–149
Ethylene
  competitive bromination, 106–108*t*
  Ziegler bromination, 108–110
Extremely toxic substances, list, 160

F

Feedstock(s), benign by design chemistry, 12–15*f*
Feedstock cost, importance in chemical manufacturing, 7
Free-radical additions and cyclizations, Mn(III)-mediated, 84–94
Free-radical reactions using supercritical carbon dioxide as medium, 98–112
  bromination of alkylaromatics, 103,106
  bromination of cyclopropylbenzene, 110–111

Free-radical reactions using supercritical carbon dioxide as medium—*Continued*
  bromination of ethylbenzene and toluene, 106,108
Friedel–Crafts reaction, photochemical alternative, 76–82

G

D-Glucose, use in synthesis of adipic acid, 33–43
Governmental initiatives, benign by design chemistry, 16–18
Guidance and evaluation protocols, initiatives for benign by design chemistry, 17

H

Halide-free routes for synthesis of aromatic amines, 133–142
Hazardous byproduct, environmental impact, 33
Hazardous chemicals, regulation via end-of-pipe approach, 166–167
Hazardous materials management, 150
HCHC–141b, high-yield–low-waste process, 145
Hexamethylenimine, resourcing, 148
High-pressure synthesis of carbamates, elucidation of mechanism, 47
High-yield–low-waste processes, 145
Housekeeping solutions, approach to pollution prevention, 5
Hydrocyanation, use of chiral phosphinite ligands, 150
Hydrogen, nucleophilic aromatic substitution, 134–142

I

Ibuprofen, synthesis via photochemical alternative to Friedel–Crafts reaction, 77,82

INDEX

Indirect electrochemical oxidations, Mn(III) mediated, 88–93
Indirect electrochemical synthesis, 86–88
Industrial applications, benign by design chemistry
  chemistry and catalysis, keys to environmentally safer processes, 144–153
  generation of urethanes and isocyanates from amines and carbon dioxide, 122–131
  halide-free routes for production of aromatic amines, 133–142
  initiatives, 13,16
  nucleophilic aromatic substitution for hydrogen, 133–142
Industries with most toxic substance releases, 7,8f
Isocyanate generation from amines and carbon dioxide, 122–131
  carbon dioxide–amine reaction mechanism, 123–124
  chemistry, 128–129
  examples, 129,130f
  experimental objective, 123
  influencing factors, 131
  mechanism, 129,130f
  synthesis using phosgene, 122–123

L

LHASA
  assessment for computer-assisted alternative synthetic design, 172–175
  ease of use and user friendliness, 179
  synthetic prediction, 179–180
  translation of chemical knowledge into synthetic proposals, 169

M

Metal-catalyzed carbonylation of nitroaromatics to aryl carbamates, 46

Methyl O-podocarpate, synthesis using Mn(OAc)$_3$, 85
1-Methyl-3,4-dihydroisoquinoline, synthetic prediction using CAMEO, 172–173
Methylglutaronitrile, waste resourcing, 146,148
Methylisocyanate synthesis, 150–151
Methylpentamethylenediamine, 146,148
4-Methylquinoline, synthetic prediction using CAMEO, 172–173
Microbial catalysis of adipic acid from D-glucose
  alleviation of rate-limiting enzymes, 39
  catalytic direction of D-glucose into aromatic biosynthesis, 34f,35–39
  choice of *Escherichia coli* as microbial catalyst, 35
  economics, 43–44
  D-glucose to *cis,cis*-muconate biosynthesis, 39–43
  use of microbial catalyst, 33,35
Minimization, waste, processes, 145–149
Mn(II) to Mn(III), chemical reoxidation, 93
Mn(III)-mediated electrochemical oxidative free-radical cyclizations, 84–94
Mn(OAc)$_3$-based oxidative free-radical additions and cyclizations, 85–86

N

N$_2$O, resourcing, 148
Naproxen, synthesis using asymmetric catalysis, 149–150
National Science Foundation
  initiatives for benign by design chemistry, 18
  program for environmentally benign synthesis and processing, 24–27
  research in environmental chemistry, 28–29
p-Nitroaniline, synthesis, 138–142
Nitroaromatics, carbonylation, 46–62

p-Nitrochlorobenzene, substituted aromatic amine synthesis, 133,134*f*
4-Nitrodiphenylamine, retrosynthetic prediction using SYNGEN, 176–177
Nitrous oxide, production during adipic acid synthesis, 33
Nucleophilic aromatic substitution for hydrogen to produce amines, 133–141
  4-aminodiphenylamine synthesis, 134–139
  p-nitroaniline synthesis, 138–142
  p-phenylenediamine synthesis, 138–142
  reaction, 133,134*f*,135
Nylon 6,6, synthesis from adipic acid, 32
Nylon monomer processes
  chemistry, 146,147*f*
  waste resourcing, 146,148–149

O

Office of Pollution Prevention and Toxics, function, 157
Optically active drugs, trend in manufacture, 149
1,3-Oxathiane(s), use in organic synthesis, 66
Oxathiane protecting group, cleavage, 70–71
Oxidations, indirect electrochemical synthesis, 88
Oxidative free-radical additions and cyclizations, Mn(III)-mediated, 84–94
Oxidative free-radical cyclization, 85
Ozone, depletion reactions, 98–99

P

2-Pentenenitrile, resourcing, 146
p-Phenylenediamine, synthesis, 138–142
Phenyl 2-thiophenyl ketone, synthetic prediction using CAMEO, 171–172
Phosgene
  alternatives for use, 46–62
  isocyanate synthesis, 122–123
Photochemical alternative to Friedel–Crafts reaction
  chemicals studied, 76–77
  diazepam synthesis, 80–81
  doxepin synthesis, 82
  experimental procedure, 76–77
  substituted benzophenone synthesis, 78–79
Pollution
  dilution as solution, 23
  potential of new chemical substances, 156
  pounds of waste released in 1992, 3,4*t,f*
Pollution prevention
  advantages, 3,64
  computer-assisted alternative synthetic design, 166–181
  description, 35
  legislation, 5
  philosophy, 167
  role of synthetic chemist, 6,7,9
Pollution Prevention Act of 1990
  description, 157
  purpose, 167
Pollution remediation, disadvantages compared to prevention, 64
Polyisocyanates, generation from amines and carbon dioxide, 129,130*f*
Premanufacture notification substances, alternative syntheses and other source reduction opportunities, 156–164
Preparative reactions using visible light
  chemical transformation studies, 66
  cleavage
    benzyl ether protecting group, 71–73
    dithiane protecting group, 67–70
    oxathiane protecting group, 70–71
  experimental description, 65–66, 73–75
Prevention of pollution, *See* Pollution prevention
Pseudoelectrochemical transformation, *See* Preparative reactions using visible light
Public–private partnerships, benign by design chemistry, 16–18

INDEX

Q

Quinoline, retrosynthetic prediction using LHASA, 174–175

R

Reductions, indirect electrochemical synthesis, 87–88
Relatively innocuous substances, list, 161
Reoxidation, chemical, Mn(II) to Mn(III), 93
Research, benign chemistry
  alternatives for use of phosgene, 46
  mechanistic study of catalytic process for carbonylation of nitroaromatic compounds, 46–62
  microbial biocatalysis, synthesis of adipic acid from D-glucose, 32–44
  Mn(III)-mediated electrochemical oxidative free-radical cyclizations, 84–96
  photochemical alternative to Friedel–Crafts reaction, 76–82
  preparative reactions using visible light, 64–74
  pseudoelectrochemical transformation, 64–74
  supercritical carbon dioxide as medium for conducting free-radical reactions, 98–112
  UCLA styrene process, 114–119
Resourcing, waste, processes, 145–149
Ru(dppe)(CO)$_2$[C(O)NHCH(CH$_3$)$_2$]$_2$, reactivity, 54
Ru(dppe)(CO)$_2$[C(O)OCH$_3$]$_2$–p-toluidine interaction, reaction, 51–54

S

Secondary amines, reaction with carbon dioxide, 123
Social pressures, environmental preservation, 24
Solvent alternatives, benign by design chemistry, 12
Source reduction
  advantages, 164
  definition, 5,157
Source reduction reviews
  assessment procedure, 159–163
  examples, 163–164
  pilot study, 159
  requirements, 157–159
State legislatures, environmental laws incorporating pollution prevention, 5
Styrene manufacture, UCLA process, 114–119
Substituted benzophenone, synthesis via photochemical alternative to Friedel–Crafts reaction, 78–79
Supercritical carbon dioxide as medium for free-radical reactions, 98–112
  advantages, 100,111–112
  alkylaromatic bromination, 106–107
  apparatus, 104,105f
  cyclopropylbenzene free-radical bromination, 110–111
  ethylbenzene Ziegler bromination, 108–110
  ethylene competitive bromination, 106,107–108t
  experimental objectives, 101,103–106
  pressure vs. reaction rate, 100
  toluene competitive bromination, 106,107–108t
  toluene Ziegler bromination, 108–110
Supercritical fluids, properties, 99–100
SYNGEN
  assessment for computer-assisted alternative synthetic design, 174–177
  ease of use and user friendliness, 180
  synthetic prediction, 180–181
  translation of chemical knowledge into synthetic proposals, 169
Synthesis
  2-acylhydroquinones, 78–79
  adipic acid from D-glucose, 32–44

Synthesis—*Continued*
  alternative design for pollution
    prevention, 166–183
  alternative, for premanufacture
    notification substances, 156–164
  4-aminodiphenylamine, 134–139
  aromatic amines, 133–142
  bicyclo[2.2.2]oct-5-ene-2,3-dicarboxylic
    anhydride, 171
  bicyclo[3.2.1]octanone, 86
  carbaryl, 170
  4-(2-carbethoxycyclopropyl)pyridine, 171
  methyl *O*-podocarpate, 85
  1-methyl-3,4-dihydroisoquinoline, 173
  methylisocyanate, 150–151
  p-nitroaniline, 138–142
  phenyl-2-thiophenyl ketone, 172
  p-phenylenediamine, 138–142
Synthetic chemist's role
  environmental chemistry, 2
  pollution prevention, 67
Synthetic design, priority on health and
  environmental impact, 98
Synthetic efficiency, role in benign by
  design chemistry, 10
Synthetic methodology assessment for
  reduction technologies
  detailed assessment, 161–163
  examples, 163–164
  preliminary assessment, 159–161

T

Tandem oxidative cyclization, alkenes, 86
Toluene
  competitive bromination, 106–108*t*
  Ziegler bromination, 108–110
Toxic materials management, 150–151
Toxic Substances Control Act, purpose, 157
Transformation, pseudoelectrochemical,
  66–71

U

UCLA styrene process
  benzocyclobutane conversion to styrene,
    116–119
  commercial development, 119
  diazomethyltoluene transformation to
    tolylmethylene, 115
  tolylmethylene interconversion to
    benzocyclobutane and styrene, 116
Urethane generation from amines and
  carbon dioxide, 122–131
  advantages, 122
  carbon dioxide–amine reaction
    mechanism, 123–124
  carbon dioxide pressure vs. reaction
    rate, 126,127*f*,128
  chemistry, 124–125
  experimental objective, 123
  influencing factors, 131
  reaction rate related to base, 125–127*t*
  yield related to base, 125,126*t*
U.S. Environmental Protection Agency
  alternative syntheses and other source
    reduction opportunities for
    premanufacture notification
    substances, 156–164
  computer-assisted alternative synthetic
    design for pollution prevention,
    166–181
  initiatives for benign by design
    chemistry, 16–17
  research partnership in environmental
    chemistry, 29
  regulatory authority over new chemical
    substances, 156

V

Visible light, preparative reactions,
  64–75

INDEX

W

Waste(s)
  economic benefits of reduction, 3
  minimization processes, 145–149
  pounds released in 1992, 3,4t,f
  resourcing processes, 145–149
  trigger levels, 162

X

Xylenes, use in styrene manufacture, 115–119

Y

Yield, importance in chemical manufacturing, 67

Z

Ziegler bromination of ethylene and toluene, 108–110

Production: Charlotte McNaughton
Indexing: Deborah H. Steiner
Acquisition: Rhonda Bitterli
Cover design: Michelle Telschow

Printed and bound by Maple Press, York, PA

Bestsellers from ACS Books

The ACS Style Guide: A Manual for Authors and Editors
Edited by Janet S. Dodd
264 pp; clothbound ISBN 0–8412–0917–0; paperback ISBN 0–8412–0943–X

The Basics of Technical Communicating
By B. Edward Cain
ACS Professional Reference Book; 198 pp;
clothbound ISBN 0–8412–1451–4; paperback ISBN 0–8412–1452–2

Chemical Activities (student and teacher editions)
By Christie L. Borgford and Lee R. Summerlin
330 pp; spiralbound ISBN 0–8412–1417–4; teacher ed. ISBN 0–8412–1416–6

*Chemical Demonstrations: A Sourcebook for Teachers,
Volumes 1 and 2,* Second Edition
Volume 1 by Lee R. Summerlin and James L. Ealy, Jr.;
Vol. 1, 198 pp; spiralbound ISBN 0–8412–1481–6;
Volume 2 by Lee R. Summerlin, Christie L. Borgford, and Julie B. Ealy
Vol. 2, 234 pp; spiralbound ISBN 0–8412–1535–9

Chemistry and Crime: From Sherlock Holmes to Today's Courtroom
Edited by Samuel M. Gerber
135 pp; clothbound ISBN 0–8412–0784–4; paperback ISBN 0–8412–0785–2

Writing the Laboratory Notebook
By Howard M. Kanare
145 pp; clothbound ISBN 0–8412–0906–5; paperback ISBN 0–8412–0933–2

Developing a Chemical Hygiene Plan
By Jay A. Young, Warren K. Kingsley, and George H. Wahl, Jr.
paperback ISBN 0–8412–1876–5

Introduction to Microwave Sample Preparation: Theory and Practice
Edited by H. M. Kingston and Lois B. Jassie
263 pp; clothbound ISBN 0–8412–1450–6

Principles of Environmental Sampling
Edited by Lawrence H. Keith
ACS Professional Reference Book; 458 pp;
clothbound ISBN 0–8412–1173–6; paperback ISBN 0–8412–1437–9

Biotechnology and Materials Science: Chemistry for the Future
Edited by Mary L. Good (Jacqueline K. Barton, Associate Editor)
135 pp; clothbound ISBN 0–8412–1472–7; paperback ISBN 0–8412–1473–5

For further information and a free catalog of ACS books, contact:
American Chemical Society
Distribution Office, Department 225
1155 16th Street, NW, Washington, DC 20036
Telephone 800–227–5558

TP 247 .B395 1994 c.1

Benign by design